CLARENDON LIBRARY OF LOGIC AND PHILOSOPHY

General Editor: L. Jonathan Cohen

PSYCHOLOGICAL MODELS AND NEURAL MECHANISMS

Also published in this series

The Probable and Provable by L. Jonathan Cohen
Relative Identity by Nicholas Griffin
Equality, Liberty and Perfectionism by Vinit Haksar
Experiences: An Inquiry into some Ambiguities by J. M. Hinton
Knowledge by Keith Lehrer
Metaphysics and the Mind-Body Problem by Michael E. Levin
The Cement of the Universe: A Study of Causation by J. L. Mackie
Truth, Probability and Paradox by J. L. Mackie
The Nature of Necessity by Alvin Plantinga
Divine Commands and Moral Requirements by P. L. Quinn
Simplicity by Elliot Sober
The Coherence of Theism by Richard Swinburne
The Emergence of Norms by Edna Ullmann-Margalit
Ignorance: A Case for Scepticism by Peter Unger

PSYCHOLOGICAL MODELS AND NEURAL MECHANISMS

An Examination of Reductionism in Psychology

BY

AUSTEN CLARK

CLARENDON PRESS · OXFORD
1980

Oxford University Press, Walton Street, Oxford OX2 6DP

OXFORD LONDON GLASGOW
NEW YORK TORONTO MELBOURNE WELLINGTON
KUALA LUMPUR SINGAPORE HONG KONG TOKYO
DELHI BOMBAY CALCUTTA MADRAS KARACHI
NAIROBI DAR ES SALAAM CAPE TOWN

Published in the United States
by Oxford University Press
New York

British Library Cataloguing in Publication Data

Clark, Austen
Psychological models and neural mechanisms. –
(Clarendon library of logic and philosophy).
1. Psychology – Simulation methods
2. Neuropsychology
I. Title
612′.8′0184 BF38.5 79-41732

ISBN 0-19-824422-3

Set in IBM Baskerville by Graphic Services Oxford
Printed in Great Britain by
Billings and Sons Ltd.,
Guildford, London, Oxford and Worcester

Acknowledgements

Many people helped to make this book possible. I thank the Executive Secretary and Officers of the Marshall Aid Commemoration Commission for a scholarship to Queen's College, Oxford, where I wrote the thesis on which this book is based. Special thanks go to Miss Geraldine Cully, the Assistant to the Executive Secretary. I owe much to my supervisor, Mr. Brian Farrell of Corpus Christi, Oxford, for the expert supervision he gave me, and for the invaluable suggestions, comments, and criticisms he made on all the drafts of the thesis. I gained much from his knowledge of psychology and philosophy. Thanks go to Dr. E. T. Rolls for providing tutorial supervision in physiological psychology, and to Dr. G. S. Claridge and Dr. I. A. Wilson for their advice on parts of the thesis. I thank the Thetford Historical Society and Latham Memorial Library for providing me with a warm place to revise the thesis during Vermont's snowy winter of 1977–8. Finally, I am grateful to my wife, Denise Polivy, for her comments and suggestions in editing the thesis, and for her warm encouragement and support during its revisions. I dedicate this book to her.

Contents

List of Figures

Introduction

Reductionism has been a live issue since science began producing theories with implications about human behaviour. Many people are disturbed by the provision of new systems of explanation, of scientific origin, for behaviour which is traditionally understood in different terms. Such theories often seem to carry the implication that certain human traits are not what we thought they were and can be assimilated to a radically different set of traits or processes. Reductionist theories often seem to reduce human beings to something simpler than we suspected, and show that behaviour which had traditionally been assigned a certain origin and meaning has simpler, unexpected antecedents, and that the meanings traditionally assigned to them are therefore delusory.

Accounts which link certain aspects of human behaviour to supposed innate aggressive instincts and the ethological principles governing them, or to certain simple principles of reinforcement and operant conditioning, or to supposed innate behavioural tendencies to perpetuate one's genotype, are all popularly labelled 'reductionist'.[1] They all provide simple accounts of behaviour which is traditionally explained in other ways, and they assimilate that behaviour to previously unsuspected mechanisms. Such accounts are also thought to rule out the applicability of our ordinary explanations of behaviour and show that those behaviours do not have the meanings we thought they did. In this way reductionist theories seem to change irrevocably the way we describe and understand those actions.

The problem of reductionism in psychology is generated by the provision of physiological explanations for behaviour which previously had received only psychological explanations.

[1] See Konrad Lorenz, *On Aggression* (New York: Harcourt, Brace & World, 1966); B. F. Skinner, *Science and Human Behaviour* (New York: Macmillan Co., 1953); Richard Dawkins, *The Selfish Gene* (New York: Oxford University Press, 1976).

It is a truism to say that a person's psychological functioning is in some way related to what goes on in his or her nervous system, and with increasing knowledge of this relation we may be able to give neurophysiological accounts of behaviour where previously our ignorance confined us to explanations in terms of psychological states such as desires, intentions, or beliefs. The issue arises in several different ways. Psychological models developed in the experimental discipline are thought to be related in some way to underlying neural mechanisms, and as knowledge of this relation increases, perhaps a purely physiological explanation will come to replace our psychological accounts. In abnormal psychology arguments continue to rage over the supposed antagonism between a medical model of schizophrenia or depression and a learning theory account. Physiological explanations of psychosis in terms of a physiological disease process or an underlying biophysical abnormality are treated as rivals of accounts which mention a learning history, principles of learning uncovered in the laboratory, and acquired habits. The problem of reductionism is perhaps most pressing in abnormal psychology, for there the success or failure of physiological explanations of abnormality is thought to have enormous consequences on our theories, diagnosis, and therapy of people in mental hospitals and clinics.

Reductionist theories in psychology are disturbing not only because they are rivals to traditional schemes of understanding people, but because they seem to discredit or invalidate that entire class of explanation. The elucidation of simple causal mechanisms in which the antecedents of behaviour lie outside the person's cognizance or control seems to imply that desires, reasoning, and choice have nothing to do with behaviour. By providing causal antecedents from which the behaviour in question necessarily follows, we seem to show that the person could not have done anything other than what he did, and that free will is therefore a delusion.

If reductionist theories did carry the implication that reasoning and choice are of no relevance for certain behaviours, then such theories would rule out our ordinary ways of explaining such behaviours. It would be shown that reference to the person's desires, beliefs, or goals is useless in trying to explain the behaviour, and that it must be explained in terms of other causal antecedents. Reductionist theories in this way seem antagonistic to explanations in terms of reasons, so that as causal accounts succeed in explaining more and more of human behaviour, our traditional understanding of people must be progressively dismantled and abandoned.

A recurrent popular image is that of behavioural science gradually encroaching on the domain of free will, breaking off and absorbing portions of that territory as scientists find causal explanations of mental illness, deviance, and criminality; that the province of free will and action, to which our ordinary explanations validly apply, must continually shrink under the impact of increased knowledge, finally to disappear altogether with the success of physiological reductionism. When that day comes, our ordinary concepts of choice, reasons, and responsibility must be finally abandoned as superstition; for it will have been shown that what we now call 'choice' never really existed. Success of the reductionist programme is thought to invalidate that entire system of explanations; our appeal to the reasons for actions will be eliminated, replaced by reference to the physiological causes from which our behaviour ineluctably follows.

The thought that this might occur is the source of many of the worries which people have concerning an expanding scientific account of human beings. For along with the enormous changes such an account would force in the way we describe, understand, and evaluate people, it is thought to have frightening practical consequences as well. Once traditional understandings of humans are invalidated and abandoned, the door is open to increasing technological reorganization of our social institutions. Our moral notions and our system of justice, which are based on the assumption of personal responsibility, would need radical overhaul. Schools, prisons, mental hospitals, and other institutions in which participants' behaviour is changed would need to be revamped to make way for the scientific manipulation of behaviour through psycho-technology and new therapies. Our political arrangements and institutions for public policy would give way to the scientific calculation of contingencies and the institution of controlled manipulations by those expert in the abstruse theories of human behaviour. A recognition that human freedom and dignity are delusory would destroy our superstitious understandings of ourselves and facilitate the change to a totally-administered, scientifically-organized society. Granted, our contemporary understandings of people and social institutions are still largely framed in traditional terms, but their ultimate replacement is thought to be an inevitable consequence of the growth of behavioural science.

In this book I shall argue that reductionism, so conceived, is a bogeyman. By this I mean not that the fears it engenders are baseless, but that they are misplaced. There is much that is worrisome about some of the ways in which theories of human behaviour are applied in some mental hospitals, schools, and prisons, but I shall argue that it is a mistake to think that the reduction of psychological explanations to physiological once entails the above consequences. To do this I shall present an analysis of the necessary and sufficient conditions for reducing a psychological model to physiology, and then consider whether such a reduction entails an incompatibility of psychological explanations. The bogeyman of reductionism relies on confusing the issue of explanatory reduction with other issues, and if we can disentangle some of these different questions, we can gain a clear view of the meanings and consequences of reducing psychological models to physiology. I hope that this enterprise will also clarify the terms of the debate between protagonists of the medical model and its opponents, and that it will help to distinguish the real issues from the unimportant ones for someone concerned about the practical applications of knowledge in behavioural science.

I
Explanatory Reduction and the Incompatibility of Explanations

1. *A Preliminary Definition*

To make a start, we need a rough definition of what 'reductionism' means. 'Explanatory reduction' is the technical term for 'reductionism', and as it has fewer emotional connotations than the latter word, I shall generally use it rather than the word 'reductionism'. As a preliminary characterization, claims about explanatory reduction are claims that a particular *relation* obtains between two *theories* in different disciplines. There are different analyses—which I shall discuss below—of exactly what sort of relation is necessary and sufficient for explanatory reduction to succeed, and some philosophers contend that the terms of this relation involve models and not theories, but these details are inessential to the main point. When we say that 'psychology reduces to physiology' we mean (roughly) that some physiological theory can be used to explain everything that is explained by some psychological theory. Strictly speaking, it is incorrect to say that 'psychology reduces to physiology', for explanatory reduction is not a claim about two disciplines *per se*; rather it is a claim about *theories* from two disciplines. To ask a question about explanatory reduction meaningfully, therefore, we must first specify which two theories we are considering.

This is a very simple point, but it can go far towards ending some of the confusions one finds in arguments about reductionism. Often claims about the relations between theories in two disciplines are confused with claims about other features of the two disciplines, and in this way explanatory reduction is confused with other issues. For example, sometimes it is thought that if a physiological account of schizophrenia were given, the problem of schizophrenia would thereby be relegated to physiology, and that thereafter psychologists could have little to contribute in terms of research or therapies. It is thought that psychological research into behavioural problems is provisional upon the absence of a physiological account, for once such an account is successful, research must be handed over to the physiologists and therapy design to psychiatrists. Yet surely this claim confuses questions about theories with questions about appropriate strategies of research and the appropriate field from which to draw therapies. It is analogous to claiming that because the thermodynamic gas laws reduce to theories in statistical mechanics, thermodynamic research into gases is thereafter inappropriate, and our calculations for designing furnaces and radiators must thereafter be made in statistical mechanical terms! Demonstrating a relation between two theories does not necessarily show anything about appropriate strategies of research or the practical application of those theories.

One connotation which is carried by the term 'reductionism', but not by the term 'explanatory reduction', is that the new account of behaviour is much simpler than the old one, thereby showing that reference to the more complicated states and processes invoked by the old account is unnecessary in explaining the behaviour. Reductionist theories are thought to *reduce* human behaviour to something simpler by demonstrating that our more complex explanations are unnecessary. The term 'explanatory reduction', however, does not imply that the reducing theory is necessarily simpler than the reduced theory. In fact, the reducing theory will generally be *more* complicated than the reduced theory in any case of micro-reduction, in which we explain the observable states and processes of some system by citing micro-constituents ('cells', 'atoms', and so on) with certain properties, interacting in certain ways. To think explanatory reduction entails the success of simpler explanations confuses claims about the relations between theories with substantive claims about the *content* of the theories.

Explanations which cite a system's micro-constituents sometimes lead to another confusion. It is very commonly thought that if some system *x* is composed of parts *y*, then laws about the behaviour of *x* must be reducible to the laws concerning the constituents *y*. For example, in discussing whether or not the brain is a physical system, N. S. Sutherland writes:

Our claim is not that the explanation of behaviour does not involve new principles but only that if the brain is a physical system then these new principles are reducible to the laws governing the behaviour of elementary particles.[1]

But this plausible claim conflates the distinction between a theory and the objects the theory is about. The fact that a system *x* is composed of parts *y* does not entail that *theories* about *x* are related in some particular way to *theories* about *y*. We may know that *x* is composed of parts *y* yet have no idea of how those parts work together to produce the observable phenomena. Since in that case we have no theories about *y*, our theories about *x* are not reducible to theories about *y*. For example, even though neurons are composed of molecules, which in turn are composed of elementary particles, even a molecular explanation of how the cells work is as yet incomplete. For this reason we cannot reduce a physiological explanation of neural functioning to a molecular explanation. Likewise, even though we know that the brain is composed of neurons, in the absence of a physiological theory of behaviour we cannot say that psychological explanations of behaviour reduce to physiological ones.

Perhaps Sutherland means to claim that there can (or will) be some future physiological theory of behaviour to which our psychological explanations will reduce. Claims concerning the ultimate reducibility of one future theory to another future theory can sometimes be construed as recommendations that research pursue one course rather than another. For example, one might propose that research shift its focus from discovery of observable regularities in learning to a search for their underlying physiological mechanisms. Such claims can also be construed as denials of the various arguments that it is logically impossible that there ever be some physiological theory to

[1] N. S. Sutherland, 'Is the brain a physical system?', in *Explanation in the Behavioural Sciences*, eds. Robert Borger and Frank Cioffi (Cambridge: Cambridge University Press, 1970).

which our psychological explanations of behaviour reduce. Sutherland is in fact making the latter sort of claim. In either case, claims asserting or denying the *reducibility* of one class of explanations to another are distinct from claims to the effect that one particular theory T does in fact *reduce* to another particular theory T'.

The distinction between theories and their objects has a further consequence. Ernest Nagel writes that in explanatory reduction:

> a set of distinctive traits of some subject matter is assimilated to what is patently a set of quite dissimilar traits. In such cases, the distinctive traits that are the subject matter of the secondary science fall into the province of a theory that may have been designed for handling qualitatively different materials and that does not even include some of the characteristic descriptive terms of the secondary science in its own set of basic theoretical distinctions. The primary science thus seems to wipe out familiar distinctions as spurious, and appears to maintain that what are prima facie indisputably different traits of things are really identical.[2]

As Nagel proceeds to explain, explanatory reduction is a claim about theories, and not about the traits or characteristics ascribed by those theories. To claim that one theory reduces to another does not entail that the characteristics ascribed by one theory are illusory, or that their ultimate reality consists of the constituents and characteristics invoked by the reducing theory. Temperature is a property of gases even though the gas laws have been reduced to statistical mechanics. If explanatory reduction is successful, we have a second theory which explains everything explained by some other theory. Such a claim does not imply that the traits or processes ascribed by the reduced theory do not exist!

2. *The Incompatibility of Explanations*

Not all of the arguments about the prospect of reductionism in psychology can be dispelled by paying attention to some simple logical distinctions. Even after explanatory reduction has been carefully distinguished from questions about the strategies of research, the appropriate source of therapies, the

[2] Ernest Nagel, *The Structure of Science* (London: Routledge & Kegan Paul, 1961), p. 340. See also 361–6 for a lucid analysis of some basic issues in the logic of reduction.

simplicity of human behaviour, the ultimate constituents of nature, the purported ultimate reducibility of one discipline to another, and the reality of properties—the reduction of psychological to physiological explanations faces further difficulties.

The central fear associated with reductionism is that physiological explanations are incompatible with our current psychological accounts, so that the provision of a physiological account would entail an invalidation and elimination of our psychological explanations. To reduce some psychological account of behaviour to a physiological explanation is thought to rule out the applicability of psychological accounts for that behaviour; a physiological reductionism purportedly reveals the basic, underlying physiological antecedents of the behaviour, thereby showing that accounts which link the behaviour to psychological factors are simply incorrect.

By the *incompatibility* of explanations I mean the claim that psychological accounts and physiological accounts cannot both apply to the same behaviour, so that if one of them provides an adequate account of it, then the other is ruled out. Reductionism poses a problem because it seems impossible that both disciplines could provide adequate accounts of a given behaviour. Perhaps the best way to explore this idea of an incompatibility of explanations is to consider an example of some behavioural phenomenon for which physiological accounts have been proposed. Reactive depression is one such example. I shall sketch the outline of the emerging physiological account of reactive depression and then analyse some of the seeming implications of that account.

A. *The Example of Reactive Depression*

The central feature of reactive depression is that the depressed person feels that there is no point in continuing to live—that there is no goal worth seeking, or that even though there may be some goals worth seeking, he or she cannot attain them, and so should not even try. The future is devoid of anything valuable or interesting; no activity seems better than another because all are equally without pleasure. Along with these motivational deficits, the depressed person experiences cognitive changes: he loses his ability successfully to attain ends he desires; he has faulty perceptions about how to attain goals, a limited and repetitious thought content, and a massive collection of negative self-appraisals usually centring on loss of competence and effectiveness. He has difficulty concentrating and shows decreased alertness. He feels useless, unworthy, inadequate, or incompetent, and tends to blame himself for the depression, thinking of it as a punishment visited upon him for his sins or his failings.

These points will be briefly expanded. At a very simple level, the motivational deficit in depression consists of loss of the ability to attain normally-experienced pleasures or satisfactions in everyday activities. The depressed person loses, to an extent, his former interests, his appetites, and his sexual desires. He is not pushed or pulled by pleasures or pains as is the ordinary mortal, but is apathetic and detached. Two symptoms commonly accompany this syndrome. First, the depressed person loses energy and initiative, and complains of being fatigued. Second, he becomes withdrawn and inactive, dropping many of his former activities and restricting himself to a predictable daily routine.

These basic symptons are sometimes accompanied by others such as sleep difficulties, diurnal variation in mood, autonomic complaints, and angry, complaining, provocative behaviour alternating with depressed withdrawal.

There are grounds for thinking that the underlying physiological basis of depression may soon be understood. Many of the symptoms of depression can plausibly be construed as products of alteration in the function of certain structures in the nervous system. I do not intend to explore the details of the experimental findings or to propose a model which explains all of the symptoms (none exists, although some unverified models explain most of the symptoms). Instead I intend merely to give an example of what such a model may look like, and of the sorts of considerations involved in reducing it to neural mechanisms.

This model is based on Martin Seligman's idea of learned helplessness.[3] According to Seligman the basic psychological deficit in depression is that reinforcement no longer produces

[3] Martin Seligman, *Helplessness* (San Francisco: W. H. Freeman & Co., 1975).

the changes in behaviour that it does when a person is undepressed. Withdrawal, apathy, inactivity, loss of initiative, and fatigue are all thought to be results of the fact that behaviour is no longer effectively reinforced. The person ceases to act because he is no longer rewarded for acting; his complaints that nothing is desirable or interesting are symptoms of the underlying fact that he does not experience enough pleasure—or more precisely, that reinforcers have lost some of their reinforcing effect.

The origin of this condition is a learning history of uncontrollable aversive stimuli. If, no matter what he does, the person receives punishing stimuli—or loses rewarding ones—then he learns that his instrumental responding does not prevent the punishment, and so he ceases responding instrumentally. The death of a loved one or failure in some important project are simple examples. The withdrawal this naturally produces leads to further loss of reinforcement, and as instrumental responding extinguishes, the person gradually loses the ability to obtain the reinforcement he seeks; he expects uncontrollable punishment (or loss of reward) and so ceases responding. The basic expectation that actions can make no difference to environmental contingencies of reinforcement becomes a self-fulfilling prophecy.

The expectation of uncontrollable aversive stimuli is assumed to lead to further effects. Emotionally it produces anxiety, fear, and difficulties in interpreting the motivational significance of ordinary stimuli. The expectation that responding makes no difference leads to decreased incentive to act, decreased drive and alertness, and a drop in instrumental responding. Cognitively it produces an inability to perceive success when attained, a breakdown of discrimination learning, a negative cognitive set, and generally, a loss of skill in attaining reinforcers. These various changes all result in reinforcers having decreased effects on behaviour.

This model bears interesting relations to the catecholamine hypothesis of depressive disorders: the idea that depression is correlated with a relative deficiency of the catecholamine transmitter substances at synapses in the limbic system.[4] The

[4] J. J. Schildkraut, 'Neuropharmacology of the affective disorders', *Annual Review of Pharmacology*, 13 (1973), 427–54.

limbic system comprises various subcortical nuclei (amygdala, hippocampus, septum, preoptic area, and parts of the hypothalamus), the fibre tracts connecting them, parts of primitive cortex (temporal lobe, orbito-frontal cortex, cingulate gyrus, hippocampal gyrus, pyriform lobe), and the reticular formation and associated structures (midbrain tegmentum, intralaminar thalamic nuclei). There is very strong evidence that the limbic system is the physiological basis for many functions of motivation and emotion, and so it is plausible to assume that depression involves a change in limbic functioning.

First, and most importantly, the systems for registering reward are primarily limbic. Dopaminergic and noradrenergic fibre tracts in the limbic system are the most significant sites for brain stimulation reward. Cells in these regions fire when the animal encounters a natural reward such as food.[5] If reward depends upon the firing of certain cells in the medial forebrain bundle or the lateral hypothalamus, then a relative deficiency of transmitter substance in these structures could produce a decrease in reward. Since it seems likely that depression is associated with a deficiency of noradrenaline in limbic sites, and that a change in the parameters of reinforcement underlies the central symptoms of depression, it seems plausible to claim that the physiological basis of depression is a malfunction in the mechanisms of reinforcement; specifically, a deficiency of transmitter substance. This hypothesis has been advanced.[6]

Since limbic structures also mediate primary drives such as hunger and sex, a deficit of transmitter substance there could also account for the loss of appetite and libido in cases of depression, and for the generally decreased level of drive.

The reticular formation mediates arousal, and various parts of the limbic system are crucial in regulating alertness, attention, habituation, and the sleep-waking cycle. A deficiency of transmitter substance in these structures could conceivably result in the decreased alertness, fatigue, and impaired vitality in depression. An inability to concentrate resulting from malfunctions in attentional mechanisms could be another result.

[5] E. T. Rolls, *The Brain and Reward* (Oxford: Pergamon, 1975), pp. 19–21.

[6] H. S. Akiskal and W. T. McKinney, Jr., 'Depressive disorders: toward a unified hypothesis', *Science*, 182 (1973), 20–9.

Since reinforcement is involved with learning, we should expect that limbic structures will be involved in learning and that there is a learning deficit in depression. Both of these expectations are fulfilled. The limbic system is involved in an animal's ability to learn to associate rewards with previously neutral stimuli. This sort of learning involves functions such as (1) linking particular classes of stimuli with particular rewards often associated with them; (2) using such stimulus-reinforcement associations to interpret the significance of current stimuli; (3) on that basis arousing and orienting the animal in cases of significant stimuli, and allowing habituation in cases of unimportant ones; (4) adjusting response hierarchies and strategies to reflect current stimulus-reinforcement associations; (5) discarding ('unlearning') associations in which the correlation between stimulus and reinforcer ceases to be significant; and (6) adjusting thresholds of irritability and reactivity to stimuli accordingly. Experiments which combine lesions or stimulation of various limbic nuclei with classical-conditioning trials provide strong evidence that different limbic structures contribute in different ways to all of the above jobs.[7]

If depression involves a temporary change in the parameters of reinforcement then we would expect there to be greater difficulty learning instrumental responses. In fact depressed patients do give up on rather simple tasks, assuming that they cannot learn how to perform them and that they will always be ineffective in getting what they want. The expectation of uncontrollability leads to learning deficits.[8] A malfunction in the system associating positive or negative reinforcers with stimuli would interfere with the ability to predict the reinforcing contingencies of actions. An inability to discriminate adequately between stimuli with different associated reinforcers would also be expected of such a physiological condition—and may go some way towards explaining the ineffectiveness of depressed people and their inability to perceive success when attained. Perhaps the person misinterprets the motivational significance of ordinary stimuli.

Finally, the catecholamine systems mediate the transition

[7] See J. R. Smythies, *Brain Mechanisms and Behaviour,* 2nd edn. (Oxford: Blackwell Scientific Publications, 1970), Chs. 3–8.

[8] Seligman, *Helplessness,* pp. 87–8.

from waking to slow wave sleep, and a malfunction in those limbic structures could account for the sleep disorder found in depression. Various autonomic functions are carried out by the hypothalamus, and alteration of hypothalamic activity could account for some of the autonomic complaints of depressed people.

In summary, the limbic system has been implicated in processes of drive, reinforcement, arousal and attention, classical conditioning of reinforcers to neutral stimuli, and various autonomic functions. A malfunction in this system could account for many of the symptoms of motivational deficit in depression, and given that depressed patients apparently display a relative deficiency of certain transmitter substances in limbic sites, a model for depression relating the symptoms to underlying physiological changes can be developed.[9] Of course many of the symptoms remain unexplained and unmentioned by this model, such as the cognitive deficits in depression, the feelings of guilt and sinfulness, and the mood of worthlessness and pointlessness. Nevertheless the example may suffice to reveal some of the supposed incompatibilities of psychological and physiological accounts.

B. *The Incompatibility of Psychological and Physiological Explanations*

Why should we suppose that Seligman's learned helplessness model of depression is incompatible with a physiological account? The argument will be presented in three stages, each more difficult to answer than the last.

(1) It is apparent that the two explanations employ concepts of different kinds. Seligman's model employs concepts which are tied to behavioural contexts. Depression is characterized as a condition of helplessness and ineffectiveness in which much of a person's instrumental behaviour extinguishes because he or she expects that behaviour makes no difference to the positive or negative reinforcers received. An 'expectation of uncontrollability' leads to 'motivational' and 'cognitive' deficits. The concepts of 'helplessness', 'ineffectiveness', 'expectation', 'extinction of instrumental behaviour', and 'cogni-

[9] Smythies, *Brain Mechanisms and Behaviour*, pp. 158–69.

tive deficit' are all psychological concepts, describing dispositions to behave in certain ways in certain situations. They are all terms whose application is determined by the presence of certain behaviours in certain situations.

Furthermore, explanations employing these concepts proceed in a particular way. Depression is explained by citing certain stimulus events and their effect on various psychological processes, which in turn lead to the depressed syndrome. The antecedents cited all have to do with features of the person's learning history, and their relation to depression is explained by describing their effect on emotional, motivational, and cognitive processes. In particular, certain contingencies lead to an expectation of uncontrollability, which in turn results in an inability to perceive success, misinterpretation of stimuli, and decreased effectiveness of reinforcers. Depression is identified as a behavioural syndrome (a constellation of dispositions, incapacities, inclinations, and so on) and is explained by showing the effect of a particular learning history on psychological processes.

In contrast, a physiological model of depression seems to operate with entirely different concepts and observations and to proceed in a different way. Depression is identified as an abnormality in certain structures in the limbic system: a relative deficiency of transmitter substance in certain mechanisms. The concepts used to describe this condition have none of the links to behaviour which are found in psychological concepts. Instead we describe the depression in terms of decreased conductivity in the reward mechanism, a malfunction in limbic structures responsible for linking reinforcers to neutral stimuli, an alteration in function of the reticular formation, and so on. The physiological concepts describe various structures in the nervous system and their jobs. For example, decreased effectiveness of reinforcers might be redescribed as a change in the parameters of function of the mechanisms of reward in the limbic system. Instead of describing the emotional concomitants of depression, one describes a malfunction in the limbic structures responsible for classical conditioning of reinforcers to initially neutral stimuli.

Physiological explanations of depression seem to proceed in a different way from psychological ones. Instead of referring to a person's learning history and the effect of non-contingent punishments on motivational and cognitive processes, the physiological explanation refers to the physiological causes of a deficiency of transmitter substance in the limbic system. Certain events in the person's history lead to a temporary functional derangement of diencephalic mechanisms of reinforcement, which in turn causes further functional changes in the related limbic mechanisms controlling drive, arousal and alertness, sleep, classical conditioning, and autonomic activity.[10]

The observations cited in physiological explanations are in a different class from those of psychological explanations, for they require physiological assay or measurement, rather than reference to events in a learning history. The physiological observations are linked to the depressive syndrome not by their effect on psychological processes (expectation, cognition, and motivation) but by the effects they produce in limbic functioning. For example, instead of explaining the decreased incentive to act by relating it to a motivational change produced by an expectation of uncontrollability, one redescribes the 'decreased incentive' as a malfunction in the mechanisms underlying drive, reinforcement, and arousal. That malfunction is in turn explained as an effect of the relative deficiency of noradrenaline in parts of the limbic system.[11]

In all of these ways psychological and physiological accounts of depression derive from different conceptual frameworks: the two explanations employ concepts of different kinds, cite different sorts of observations and antecedents, and refer to different sorts of intervening processes. The two accounts seem to be not merely rival theories, but also products of entirely different theoretical frameworks. They assimilate depression to different observations and different processes and in that way seem to be explaining different phenomena when they explain depression. One explains a learned behavioural syndrome while the other explains malfunction in limbic mechanisms.

It seems impossible that both accounts could validly apply

[10] See Akiskal and McKinney, 'Depressive disorders'.

[11] Smythies speculates that depressive delusions originate from physiological malfunctions in amygdala and hippocampal circuits involved in associating reinforcers with previously neutral stimuli. See *Brain Mechanisms and Behaviour*, p. 161.

to the same phenomenon. If we mean by 'depression' a psychological syndrome marked by absence of motivation, feelings of worthlessness, negative cognitive set, and so on, then reference to underlying physiological mechanisms, changed parameters, or malfunctions seems irrelevant to the understanding we seek. Those physiological terms do not explain the *psychological* syndrome. Given the success of our psychological concepts in explaining depression, it seems impossible to assimilate it to a radically different conceptual framework.

Conversely, if the physiological account is successful in explaining depression, it would seem to show that depression is not what we thought it was, that it is just a malfunction in certain limbic structures, and our psychological concepts would seemingly no longer apply. Reference to learning history, changed incentives, cognitive set, and so on, is not needed to explain depression, for it is just a malfunction in neural mechanisms. This would seem to entail that we must abandon our psychological explanations of depression and replace them with accounts referring to processes and mechanisms of the limbic system. The differences between the explanations render them incompatible: if one is successful then the other is ruled out.

(2) Considerations of scientific parsimony and theoretic strength suggest that physiological accounts will be the ultimate victors in this rivalry. If reduction were successful, physiological explanations would be more precise than psychological ones, for instead of referring to motivational and cognitive processes related in no clear way to observables, we would refer to identifiable states and mechanisms of the nervous system. Our physiological explanations would also be of wider scope and greater generality, for they could explain normal mood swings as well as the pathological variants, psychopharmacological and biochemical data untouched by learning theory accounts, depressions which are endogenous as well as those related to environmental events, and some animal learning phenomena as well as human cases. The physiological explanation would give us a precise idea of the variables which determine why some people get depressed in certain situations while others do not; and of what sorts of situations, in what severity, lead to reactive depressions, and which do not. This

greater generality of the physiological explanation would allow more indirect data to bear on testing the hypothesis, and it would subsume previously unrelated phenomena under explanatory laws.

It is a canon of scientific method to prefer a hypothesis which has greater precision, wider scope, greater explanatory power, and more supporting evidence over one which does poorly in these ways. For these reasons it seems the more scientific course to accept the physiological account of depression and abandon the psychological one. Since the physiological account explains everything our psychological account explained, and in a simpler and more powerful way, we can simply abandon the psychological theory and replace it with a physiological account. We shall no longer mention complicated and unnecessary psychological variables in our accounts of depression. That entire theoretical framework will be scrapped, and those terms will pass out of use. Just as 'phlogiston' is no longer invoked in explanations of heat phenomena, so the various psychological terms will drop out of use in explanations of depression.

(3) Indeed, should physiological accounts of depression succeed, this sort of elimination and replacement of psychological accounts is not merely optional—and accepted for reasons of parsimony and strength—but seems forced upon us. Reduction seems to force revisions in our idea of what depression is and what its antecedents are.

Our psychological accounts presuppose that various behavioural and psychological factors—non-contingent punishments, an expectation of uncontrollability, a decreased effect of reinforcers, a negative cognitive set—play a role in the genesis of the depression. If, however, a physiological account of depression is successful, it proceeds without mentioning any of those psychological variables: it does not mention contingencies of reinforcement, decreased effectiveness of reinforcers, expectations, cognitions, or attitudes. It mentions only various mechanisms in the limbic system, and the impact of certain events on their functioning. That a purely physiological account of depression is successful seems to show that the psychological variables are of no importance to the genesis of depression.

There certainly are cases in which finding a physiological cause of the depression rules out explanations which refer to psychological factors. For example, patients recovering from heart attacks are sometimes given reserpine, a drug which reduces blood pressure. One of its side-effects is to deplete the brain of noradrenaline, and in a sizeable percentage of such cases the patient develops a syndrome which is clinically indistinguishable from severe depression.[12] He feels helpless and ineffective, draws little enjoyment from his daily activities, and comes to think he will never achieve anything he wants. In this case the explanation in terms of a learning history, an expectation of uncontrollability, and its various cognitive and motivational effects, is ruled inapplicable. Although one may find these psychological factors present in reserpine depression, they are not factors relevant to its genesis, but merely further *symptoms* of the relative deficiency of noradrenaline in limbic sites. They are the by-products of the underlying physiological change. Similarly, when we can explain a depression by citing disruption of circadian rhythms (as in trans-atlantic depression), abnormal hormonal states, or drug side-effects, we rule out explaining the depression in terms of psychological factors.

Success of physiological reductionism seems to assimilate all cases of depression to our reserpine example. If we can explain the depression in purely physiological terms, then we seem to show that psychological factors are not relevant to its genesis.

Explanatory reduction therefore does not merely show that we need not employ psychological terms in our explanations of depression and can use the simpler and more powerful physiological accounts in their place; instead it seemingly shows that those psychological accounts are incorrect, and that we must abandon them. Success of physiological reductionism will, on this line, force us to change our ideas of depression. Instead of understanding it as a behavioural syndrome, related to a certain learning history and certain cognitive and motivational changes, we must think of it as a physiological malfunction, affecting limbic mechanisms of reward, drive, arousal, and classical conditioning. The various psychological factors

[12] J. J. Schildkraut and S. S. Kety, 'Biogenic amines and emotion', *Science*, 156 (1967), 21–30.

we currently use to explain depression are not its true antecedents; they are mere by-products of the underlying physiological change. Depression is a temporary malfunction of certain brain mechanisms, caused by certain environmental events impinging upon the person, and leading to various physiological, motivational, and cognitive symptoms. The person's learning, expectations, cognitive set, incentives, or inter-personal relationships play no role in the origin of the depression; it is caused by neural conditions outside his knowledge or control. He passively and unwittingly falls victim to certain changes in his neural circuitry.

In summary, the fact that psychological and physiological explanations employ different conceptual frameworks poses a problem for the prospects of reductionism in psychology. The concepts used in the two sorts of explanations are different: the psychological concepts relate to behaviour, while physiological concepts refer to structures or processes in the nervous system. The two accounts have different descriptions of the depressive syndrome, one defining it in terms of various behavioural symptoms, the other identifying it as a temporary malfunction in certain limbic mechanisms. They seem to cite different antecedents and different intervening processes. For these reasons physiology and psychology seem to embrace two different ways of conceptualizing depression—one explaining a behavioural syndrome, and the other explaining limbic malfunction.

Because the two conceptual schemes are so radically different, the suggestion that physiological accounts could take over the explanatory jobs of our psychological explanations seems extremely mystifying. If depression can be successfully assimilated to one of the theories, then it seems the other is ruled out, for it appears impossible that the syndrome could be explained equally well by two such different conceptual frameworks. Given those differences, a successful physiological account would seem to show that we need not refer to psychological factors in order to explain depression, and that we could simply abandon our psychological conceptual scheme. More strongly, those differences seem to entail that if a physiological account were successful, then our psychological accounts would be shown to be incorrect. Success of physiological reduction is thought to show that depression does not have the psychological antecedents we think it does, and it thereby must radically change our understanding of the condition. Psychological accounts will be eliminated and replaced by physiological explanations.

The invalidation of our psychological accounts by physiology is of a radical order, akin to showing that our ordinary conceptual scheme for understanding depression is a superstition. Once we abandon the psychological theoretical framework, we withdraw commitment from the psychological factors it proposes. We accept that our psychological explanations never really explained depression, and we replace such accounts with physiological explanations mentioning different antecedents. In this way the psychological conceptual scheme for explaining depression is destroyed and commitment to the existence of various psychological factors in its origin is withdrawn. Depression would thenceforth be treated as nothing but a physiological malfunction.

C. *The Replacement Assumption*

There is an unstated assumption in the above argument, which must now be brought to the fore.

We can admit that psychological and physiological explanations of depression derive from different conceptual schemes: that they employ different concepts and cite different observations in explanations of depression. There is a prima-facie and undeniable difference between describing depression as a behavioural syndrome, and describing it as a condition of decreased conductivity in certain neural structures. The sorts of observations required to apply the terms are different, and the links to the depressive syndrome are made in different ways.

Some psychologists and philosophers have cited these differences as premisses in an argument to the effect that physiological explanations of depression are impossible. What we understand by 'depression' is a certain behavioural syndrome related to a person's expectations, goals, attitudes, and values, and a physiological explanation cannot explain those aspects of the syndrome. For example, in making such an argument concerning schizophrenia, David Cooper claims:

> Reductive analyses . . . may very competently and in detail portray the extra- and intra-organismal background against which the person stands, but in each case, and for the same reasons, the personal reality itself is omitted. . . . The passively produced point of intersection of a series of abstract theoretical lines is proposed as the (more or less) irreducible reality of the person. All this is not only beside the point when we question a personal reality, it is a point about something else entirely. . . . If the central 'schizophrenia problem' is posed in terms of persons acting in relation, then we are patently operating on a level beyond reduction to a quantitative formalism.[13]

If one consistently follows the implications of this position, one must conclude that for some actions our psychological explanations will always continue to apply, and that for them, neurophysiological explanations will never be found. If one thinks that a science of behaviour is possible, one must hold that such a science employs concepts and explanations quite distinct from those of natural science. This often motivates the thesis that the distinction between behavioural (or social) science and natural science is not merely that the former uses a different kind of theory, but that it is a different kind of science—with different explanations, different concepts, and a different methodology—from the latter. Cooper, for example, goes on to say:

> The implicit rationality of such causalistic theories, which we term analytic rationality, excludes by definition any comprehension of relations of interiority between persons (sometimes called intersubjectivity) and yet it is such relations that mediate the series of acts which we call 'schizophrenic behaviour'. . . . if we are to discover the intelligibility of schizophrenic behaviour, or any other sort of behaviour, we require not merely some special descriptive technique but a type of rationality radically distinct from the analytic rationality of natural science.[14]

One can, however, take all the differences between psychological and physiological explanations of depression in a different way, and claim that they show, not that explanatory reduction cannot or will not occur, but that when physiological explanations are found, the theoretical framework of

[13] David Cooper, *Psychiatry and Anti-Psychiatry* (Frogmore, St. Albans, Herts.: Paladin, 1970), pp. 24–5.

[14] Ibid., pp. 21–2.

psychology will be invalidated and abandoned. One assumes that the success of physiological explanations would show that depression is not what we thought it was, that our current conceptual scheme is mere superstition, and that the antecedents of depression are different from those we currently assign. B. F. Skinner takes this line. He writes:

> It is in the nature of an experimental analysis of human behaviour that it should strip away the functions previously assigned to autonomous man and transfer them one by one to the controlling environment. The analysis leaves less and less for autonomous man to do. . . . Autonomous man is a device used to explain what we cannot explain in any other way. He has been constructed from our ignorance, and as our understanding increases, the very stuff of which he is composed vanishes.[15]

Since psychological and physiological explanations are of different kinds, the success of the physiological variety seems to show our psychological accounts do not apply.

Both of these positions are responses to the differences between psychological and physiological explanations. Both positions assume that those differences imply that only one sort of explanation can be adequate for a given behaviour. Cooper assumes that the adequacy of our psychological accounts rules out the possibility of ever reducing them to physiology; while Skinner assumes that eventual success of a system of physiological explanations would rule out our current psychological explanations.

Why should the differences between the two kinds of explanation lead one to conclude that they are incompatible? Consider again the preliminary definition of reduction. Explanatory reduction occurs if and only if one theory explains everything explained by a theory in a different discipline. This seems to entail that the explanatory jobs of the reduced theory can be taken over by a theory from a different discipline. Such a replacement would entail that the vocabulary of the reduced theory is redundant, and that all its explanatory and descriptive uses can be taken over by the vocabulary of the reducing theory. We seem not to need two systems of explanation, for the reducing theory can replace the reduced theory in all its explanatory roles.

[15] B. F. Skinner, *Beyond Freedom and Dignity* (Harmondsworth, Middx.: Penguin Books, 1971), pp. 194, 196.

The assumption that explanatory reduction entails a replaceability of explanations is so important that I shall label it the *Replacement assumption.* It is:

> If a theory T' reduces to a theory T, then any explanation using a term from T' can be replaced by some explanation using a term or terms from T.

In other words, if a theory T' reduces to a theory T, then the vocabulary of T' can, in all its explanatory uses, be replaced by the vocabulary of T. Such replaceability seems to be an inherent feature of the gradual unification of various piecemeal explanations of behaviour into one coherent system of scientific explanation. Ordinary accounts are replaced by systematic theory.

The differences between psychological and physiological explanations certainly preclude replacing one sort of explanation with the other. But this entails that they cannot both be applicable to the same behaviour only if we accept the Replacement assumption. The claim that the success of our psychological explanations of behaviour precludes the applicability of any physiological account tacitly accepts this assumption. The claim that the success of physiological explanations of behaviour would invalidate and replace our psychological accounts also tacitly relies upon the Replacement assumption. In this way, the Replacement assumption is a tacit premiss both of those who think that the success of our current psychological explanations precludes discovery of physiological accounts, and of those who think that if an adequate physiological explanation of behaviour were given, it would invalidate and replace our psychological accounts.

3. *Types of Explanatory Reduction*

A general conclusion emerging from the previous section is that explanatory reduction is ordinarily thought to entail the elimination and replacement of the reduced theory. Now while such elimination and replacement is one sort of relation which can obtain between reducing and reduced theories in instances of explanatory reduction, other sorts of relations are possible. Different instances of explanatory reduction in the history of science have different features, and various relations can obtain between reducing and reduced theories. Although the above example may strongly suggest that the relation between psychological and physiological explanations will be that of elimination and replacement, it does not establish that claim, for in fact our psychological explanations of depression have not—as yet, at least—been superseded by physiological accounts. The example merely illustrates common assumptions and fears about the *future* course of inquiry. For the time being it is an open question which type of reduction, if any, will obtain between the two accounts.

In order to get a clear idea of the features of explanatory reduction in psychology, we shall need to examine some simple models which have been related to underlying physiological mechanisms. Before proceeding, however, it may be wise to get some idea of the range of options in reduction. I shall present three alternative accounts of the conditions necessary and sufficient for one theory to reduce to another. Rather than attempting to produce an analysis which applies to every instance of explanatory reduction in science, I suggest that we consider the problem to be one of distinguishing various sub-types of reduction within the general family.[16] The following analyses can be considered as descriptions of three different types of reduction, of three different relations possible between reducing and reduced theories.

A. *Eliminative-Reduction*

According to the argument in the previous section, the reduction of psychological accounts of depression (T') to physiological accounts (T) will show the following features:

(1) The reducing theory T explains all of the observations explained by the reduced theory T', and normally it explains others as well. Our physiological theory must account for all the symptoms of reactive depression, but it can also explain some behavioural, psychopharmacological, and biochemical data, and data concerning normal mood swings and endogenous depression.

(2) Furthermore, the reducing theory T is more precise,

[16] For a review of different analyses of reduction, see Kenneth F. Schaffner, 'Approaches to reduction', *Philosophy of Science*, XXXIV, 2 (June 1967), 137–47.

more general, or simpler than the reduced theory *T′*.[17] *T* may predict slightly different values for some experimental laws than does *T′*. Our physiological account would presumably refer to specific neural mechanisms, and would have wider scope than the psychological account, applying to animals as well as to humans.

(3) *T* and *T′* employ radically different theoretical frameworks. Terms within them are interrelated differently, and linked to different observations. The theoretical terms in the two theories have different meanings. It is for this reason that 'depression' seems to have different meanings in the physiological and psychological accounts. The term means different things in the two accounts because it is related to different hypothetical processes and to different observations.

(4) The success of *T* shows that some of the presuppositions of the concepts of *T′* are incorrect. For example, a successful physiological explanation of depression is thought to show that psychological factors are not relevant to depression. Theoretical terms in the two theories for this reason cannot be related to each other without either falsity or self-contradiction.[18]

(5) The success of the reducing theory *T* therefore entitles us to reject theory *T′* and cease to employ its theoretical terms. We withdraw commitment from the theoretical entities hypothesized by *T′*, and simply replace that theoretical framework with that of *T*. The theory *T′* is abandoned.

Are these features found in other instances of explanatory reduction? If so, perhaps the reduction of psychological models to physiology is an instance of a common type of reduction.

An analysis of an instance of reduction which allegedly fulfils those conditions is given by P. K. Feyerabend in his article 'Explanation, reduction, and empiricism'.[19] This example—

[17] The first two conditions are clarified in John G. Kemeny and Paul Oppenheim, 'On reduction', in *Readings in the Philosophy of Science,* ed. Baruch A. Brody (Englewood Cliffs, N.J.: Prentice-Hall, 1970).

[18] See Schaffner's account of Popper-Feyerabend-Kuhn reduction in 'Approaches to reduction'.

[19] P. K. Feyerabend, 'Explanation, reduction, and empiricism', in *Minnesota Studies in the Philosophy of Science,* Vol III, eds. Herbert Feigl and Grover Maxwell (Minneapolis: University of Minnesota Press, 1962).

the reduction of Newtonian physics to relativity theory—may clarify some of the purported features of the reduction of psychological models. According to Feyerabend, relativity theory could explain all of the data subsumed under Newtonian physics, and some unexplained anomalies as well: the perihelion of Mercury and the constant velocity of light. Relativity theory predicts slightly different values for most experimental laws, for it takes into account relativistic effects—which are negligible except at high velocities. Yet even though the predicted differences between these two theories are so slight as to be undetectable in most contexts, the theories employ radically different conceptual frameworks. The fundamental assumption of absolute space and time is replaced by entirely different relativistic assumptions. The terms 'mass', 'space', and 'time', although employed by both theories, mean very different things in the two accounts. The basic structures of theoretical assumptions and relations to observables are so different that the terms play different roles within the theories. In fact the Newtonian concepts of 'mass' or 'space' cannot be defined in terms of relativistic 'mass' or 'space', for the Newtonian concepts carry presuppositions contradicted by relativity theory—that of absolute space and time. Hence the terms change in meaning when redefined in relativity theory, and the new theoretical terms cannot in any consistent way be related to those of T'. The only ground for comparison is the relative success of the theory as a whole in accounting for observations.

However, on the grounds of the relative success of T, and its greater strength or simplicity, we are justified in abandoning the reduced theory and replacing it with the new theory. The theoretical terms of Newtonian physics drop out of use, and the hypothetical entities postulated by it are no longer accepted. Its theoretical framework is eliminated and replaced by that of relativity theory. There need not be any connections between the theoretical terms of the two theories in such reductions, for the reduced theory is rejected as a poor account of the way things are.

If any of the words employed by the reduced theory are retained—such as 'mass' or 'energy'—then, as already mentioned, reduction results in their changing in meaning. This is so because they are welded into a radically new conceptual framework, playing a role in new theoretical postulates and gaining new links to other theoretical terms and observations. The assumptions which gave the terms meaning in the reduced theory are changed, and so the meanings change. More prosaically, any new inferential connection of a theoretical term to experimental data, or to other previously-unrelated theoretical terms, changes the meaning of that theoretical term in so far as it alters the term's truth conditions and the sort of evidence that can bear on its ascription or denial. Such *meaning change* might be considered a sixth feature of eliminative-reductions.

One can find many instances of abandoned hypothetical entities in the history of science—theoretical frameworks displaced by the success of a rival account. The ether does not exist; Bohr's atom does not exist; phlogiston does not exist. The reduced theory is simply eliminated, and one adopts a view which contradicts the presuppositions of the previous theoretic concepts. The reducing theory entails that there could not be such a thing as the ether, Bohr's atom, or phlogiston. We reject the existence of the old hypothetical entities because we have found a better theory which employs a different theoretical framework. Since the reduced theory fails, there is no need to relate its theoretical terms to those of the reducing theory, and the old terms can simply drop out of use.

More dramatic examples are provided by the impact of science upon pre-scientific thinking. Zeus does not exist. Although lightning exists, there is no such thing as a thunderbolt of Zeus. The only way to capture the meanings of the eliminated explanatory terms is to say something of the form 'What we used to call "a thunderbolt of Zeus" is really nothing but an electrostatic discharge.' This makes it clear that there really is no such thing as a thunderbolt of Zeus. We do not want to say 'Thunderbolts of Zeus are electrostatic discharges', for this carries the disturbing implication that thunderbolts of Zeus exist.

Some philosophers contend that eliminative-reduction provides an accurate account of the status of our psychological terms. Such terms are purportedly theoretical terms within a framework whose continued existence is provisional upon the absence of a physiological account of behaviour. Once physiology succeeds in accounting for behaviour, however, psychological theories will be abandoned and replaced by physiological accounts, and our psychological terms will disappear. *Eliminative materialists* such as Richard Rorty or P. K. Feyerabend claim an eliminative-reduction will apply to psychological terms in general.

In order to capture the meaning of the abandoned theoretical framework, we must say something of the form 'What we used to call a "sensation" is nothing but a brain process.'[20] This form of statement is used in order to avoid commitment to the existence of sensations as now understood. Various presuppositions of our psychological concepts will be contradicted by physiological accounts, and so we shall withdraw commitment from the hypothetical entities–sensations, attitudes, desires, and so on–invoked by psychological accounts. Those terms will belong to one more abandoned theoretical framework,[21] and our talk about them will be analogous to our current talk abut the thunderbolts of Zeus or the symptoms of demonic possession. Success of the physiological account is thought to show that we could drop our psychological terms and leave our ability to describe and explain phenomena undiminished, for physiological terms could take over all those jobs. The only bar to such a replacement, according to Rorty, is the purported linguistic inconvenience involved in learning the mass of physiological terms.[22] For this reason, instead of claiming that psychological terms *will be* eliminated and replaced, once physiological accounts succeed, the Eliminative materialist is more wont to claim that psychological terms *could be* so eliminated and replaced, and *would be,* were it not so inconvenient. This inconvenience is the only barrier to an elimination and replacement of our psychological terms, discarded as parts of a defunct theoretical framework.

[20] Richard Rorty, 'Mind-body identity, privacy and categories', in *The Mind-Brain Identity Theory,* ed. C. V. Borst (London: Macmillan, 1970).

[21] See P. K. Feyerabend, 'Materialism and the mind-body problem', in *The Mind-Brain Identity Theory,* ed. Borst.

[22] Rorty, 'Mind-body identity'.

B. *Derivation-Reduction*

In some instances of reduction the reducing theory *T* does not invalidate or supersede the reduced theory *T′*. Instead *T* explains why the laws of *T′* hold. Far from replacing the explanations of the reduced theory, the reducing theory merely provides a second set of explanations for the very same phenomena.

The notion of one theory *T explaining* another theory *T′* can be analysed precisely.[23] Theories can be considered as sets of sentences useful for explaining phenomena. A law is ideally a general sentence of the form

For every *x*, if *x* is *P* then *x* is *Q*.

Such sentences[24] are useful in explaining phenomena, for descriptions of some particular antecedents ('*x* is *P*') conjoined with the law *logically imply* descriptions of other states of affairs ('*x* is *Q*'). Explanation consists of logically deriving a description of the puzzling state of affairs from a description of certain antecedents, conjoined with general laws.

Explanation in this sense is not confined to the explanation of particular states of affairs, for laws can themselves be explained by being derived from other laws. On this view, for a theory *T′* to reduce to a theory *T*, all of the sentences comprising *T′* must be *logically derived* from the sentences comprising *T*. In this way *T* explains why the laws of *T′* obtain, for they are derived from the purportedly more general laws of *T*.

Certain formal conditions must be satisfied if a theory *T* is to explain a theory *T′*.[25] First, each theory must be explicitly and precisely formulated, so that the class of sentences derivable from it can be determined. Second, the observational terms of the theory must have unambiguous meanings fixed by rules of usage or test procedures, and some theoretical

[23] Nagel, *The Structure of Science,* Ch. 11.

[24] Strictly speaking, not all theories are sets of *sentences.* If some theoretical terms are not explicitly defined, use of those terms results in theoretical formulae whose meanings are only incompletely specified. Such formulae are not sentences as normally conceived. More accurately, a theory is a *partially interpreted calculus.* On this see below pp. 128–32.

[25] Nagel, *The Structure of Science,* pp. 345–51.

terms of the theory must be sufficiently related to the observational terms so that the theory has a determinate empirical content and can confront observational data. Third, the theories must do a reasonable job of accounting for the observations and empirical regularities of the field. If these conditions are not met, we cannot settle the question of whether T explains T'. If the theories are not explicitly formulated, the logical relations between them cannot be determined. If they do not have determinate meanings, then they have no determinate empirical content, and make no determinate predictions, and so they have no use as explanations. Finally, if either theory is falsified, the attempt to explain it or use it to explain some other theory would have no point.

One of the chief problems of the reduction of psychological accounts of depression to physiology is that the two theories employ different concepts—terms with different meanings, related to different sorts of observations. The theory T employs concepts referring to processes and mechanisms of the nervous system, while the theory T' employs concepts describing behaviour or psychological states. How can T explain T' if the two theories employ different concepts? More to the point, if the theory T' employs some terms which are not found in any of the sentences of T, then it is logically impossible to derive T' from T.

For these reasons we must add to our requirements the condition that *connecting principles* relate the terms in T' which are not found in T to those of T.[26] The terms in T' must be sufficiently connected to those of T to allow a derivation to succeed; and this is done by employing connecting principles which state necessary and/or sufficient conditions for the terms in T', using the terms in T. Such connecting principles are necessary in every case of reduction in which T' employs some terms not found in T. T' in those cases cannot be derived from T alone, but must be derived from T conjoined with the connecting principles.

Nagel employs the reduction of the thermodynamic gas laws from statistical mechanics as an example.[27] The former employ such terms as 'temperature', 'pressure', and 'volume',

[26] Ibid., pp. 353–4.
[27] Ibid., pp. 342–5.

while the latter employs terms describing aggregates of molecules: their summed mass and statistical descriptions of their position, velocity, and direction of travel. How can theories from such distinct conceptual frameworks be related? One requires connecting principles relating the thermodynamic terms to those of statistical mechanics. This is done by associating 'temperature' with 'mean kinetic energy of molecules', and 'pressure' with 'force exerted by the molecules' on the walls of their container.

It should be noted that such connecting principles can take various forms. Suppose '*Q*' is a term from the reduced theory, and '*W*' is a term from the reducing theory. The connecting principles need not be biconditionals of the form

For every x, x is Q if and only if x is W.

Nor need they be identity statements (of the form $Q = W$), as in our example, where the temperature of a gas is claimed to be identical with the mean kinetic energy of its molecules. All that is required is that enough of a connection between the vocabularies be established so that the derivation succeeds. For this, connecting principles of many different forms may be adequate, such as the following:

If x is Q then x is W.
If x is Q then x is not W.
Either x is Q or x is W.

One may merely state necessary and/or sufficient conditions for terms in T' using terms in T.

Some philosophers have asserted that the connecting principles relating psychological terms to physiological terms will be identity statements. Mind-brain *Identity theorists* claim that every sensation of a particular kind P (a pain, ache, tickle, itch, or whatever) is an instance of a particular kind of brain state.[28] It is further claimed that physiologists may ultimately discover precisely which neural event ϕ is identical with a particular kind of bodily sensation. Statements of the form

Sensation of type P = Brain process ϕ

will be statements of theoretical identity, linking the theore-

[28] See the articles by U. T. Place, J. J. C. Smart, Herbert Feigl, and Thomas Nagel in *The Mind-Brain Identity Theory*, ed. Borst.

tical terms in psychology and physiology. These links act as connecting principles between psychological and physiological theories.

Herbert Feigl explicitly claims that such identity statements will function as 'bridge laws' relating terms in the two theories,[29] but most Identity theorists have not explicitly committed themselves on the issue of explanatory reduction. Therefore, when I speak of the *Identity theory* in later pages, I am referring to a view not explicitly espoused in the early articles on the subject. I think nevertheless it is an interesting possibility to consider. By the 'Identity theory' I mean the claim that the reduction of psychological models to physiology will be an instance of derivation-reduction, with general identities acting as connecting principles relating the terms in the two theories. A 'general identity' is an identity of the *classes* 'sensation of type P' and 'brain process of type ϕ', so that every instance of a sensation of type P is an instance of a brain process of type ϕ. Obviously such general identities are a far stronger sort of connecting principle than is required by derivation-reduction, but the idea of such psychophysical correlation laws—linking classes of mental events to classes of brain events—is one which has often been proposed.

C. *Model-Reduction*

To some scientists it may seem somewhat forced to claim that theories are sets of sentences, and that reduction is a matter of logical derivation. Instead it will be claimed that theoretical activity is a matter of describing hypothetical mechanisms or *models* for phenomena in some domain, and that reduction is a matter of finding a model in one discipline for a theory in another discipline.

The word 'model' has many different senses (some of which will be clarified in Chapter IV, Section 2), but in this context it denotes some system or phenomenon which is fairly well understood and which is thought to be analogous in some ways to the phenomena under investigation. The states and processes of some familiar system are thought to show the

[29] Herbert Feigl, *The 'Mental' and the 'Physical': The Essay & a Postscript* (Minneapolis: University of Minnesota Press, 1967), p. 75 and *passim.*

same structure of relationships as the hypothetical states and processes of the phenomena under investigation. In describing such a model-system for a given phenomenon, the theoretician may also be said to describe a *hypothetical mechanism* for the phenomenon: some familiar system whose internal processes are similar to those purportedly responsible for the phenomena one observes.

For example, instead of presenting a statistical mechanical account of a gas as a set of theoretical postulates related indirectly to observations, it is much more natural to present it as a *description* of a model-system thought to be analogous to a gas.[30] One describes a system of tiny particles in random motion at high velocities, whose collisions are perfectly elastic. These particles have small radii compared to the distances between them. Certain features of the model-system are related to each other in just the same way that the temperature, pressure, and volume of a gas are interrelated. Just as pressure is a kind of force, so the tiny particles colliding with the walls of their container impart force. Pressure is related to the temperature of a gas in the same way as the force imparted to the walls of the container by the tiny particles is related to their mass and velocity and to the mean number of impacts on the walls over time. Just as temperature is a form of energy, so the tiny particles have kinetic energy, which can gradually be transferred to adjacent parts.

If one proceeds statistically to characterize the mass, velocity, and direction of travel of the particles, one can logically derive equations which relate some of the features of the model-system, and which are precise analogues to the thermodynamic gas laws.[31] If one adds the assumption that under certain conditions the collisions between the particles are inelastic, one can derive an analogue for a modified gas law which better fits the experimental data.[32] Minor elaborations in the statistical description of the system of tiny particles suggest new relations of dependence between the thermo-

[30] In so far as theoretical terms are only partially defined, presentation of a theory, unlike that of a model, is not a *description*. See Nagel, *The Structure of Science*, pp. 94–5, and see below, pp. 131–3.

[31] See Berent Enç, 'Identity statements and micro-reductions', *Journal of Philosophy*, LXXIII, 11 (10 June 1976), 285–305.

[32] Nagel, *The Structure of Science*, p. 113.

dynamic gas laws and other modifications in them which bring them closer to experimental data.

To reduce the thermodynamic gas laws to statistical mechanics on this view is to find a *substantive model* in statistical mechanics for the thermodynamic gas laws.[33] The similarities between the descriptions of the model-system and the thermodynamic laws become so compelling that one ceases to treat the model-system (a collection of tiny particles) as a mere analogy for a gas. Instead one enters existence claims for the model-system: one claims that a gas *is* a collection of tiny particles, that temperature *is* the mean kinetic energy of the molecules, that pressure *is* the force exerted by those particles on the walls of their container, and so on. The model-system is a substantive model in that it is not merely analogous to a gas, but instead is claimed to be an instance of the same sort of system as the gas, with features and processes similar to those of a gas.

On this account, explanatory reduction occurs if and only if one finds a model-system M in some discipline such that terms describing features or processes of this system can be assigned to every theoretical term of a theory T' in some other discipline, and one can identify phenomena in the domain of T' as instances of the same sort of system as the model-system M. The latter condition assures that the model-system is more than a mere analogy for observed phenomena, and is claimed actually to exist.

The connecting principles relating the model-system M to the theory T' have two interesting features. The first is that they must be identity claims.[34] Model-reduction succeeds only if one finds a substantive model in one discipline for a theory in another, and this entails that one must identify certain features in the domain of the theory with features of the model-system. For example, we claim that there really do exist

[33] See Romano Harré, *The Principles of Scientific Thinking* (London: Macmillan, 1970), pp. 35–6. On models see also Roy Bhaskar, *A Realist Theory of Science* (Leeds: Leeds University Press, 1975) and J. J. C. Smart, *Philosophy and Scientific Realism* (London: Routledge & Kegan Paul, 1963).

[34] See R. Causey, 'Attribute identities in micro-reductions', *Journal of Philosophy,* LXIV, 14 (3 Aug. 1972), 407–22; Lawrence Sklar, 'Types of inter-theoretic reduction', *British Journal for the Philosophy of Science,* XVIII, 2 (Aug. 1967), 109–24.

tiny particles having the properties ascribed to the model-system, and that a gas is a collection of such tiny particles.

The second feature of such connecting principles is that they can be explained.[35] It will be recalled that in derivation-reduction the connecting principles relate terms in the two theories—and as such connecting principles are not part of either theory, they cannot be explained by either theory. In model-reduction, on the other hand, one can provide a justification for accepting the connecting principles. Starting from some basic considerations of features of the model-system, one can derive equations relating some features of the model-system to others. These equations are found to be perfectly isomorphic to the equations relating features in the domain of the theory T'.[36] This isomorphism justifies the identification of some features of the domain of T' with features of the model-system M.

For example, to explain the identity between pressure and the force the particles of the gas exert on the walls of their container, one derives some characteristics of that force from more general characteristics of the model-system. Starting from the assumptions that the particles are in random motion, have small radii compared to their inter-molecular distance, and have elastic collisions, one can derive equations describing the relation of the force those particles exert on the walls of the container to their density and mean kinetic energy. These equations are isomorphic to those describing gas pressure and its relation to volume, mass, and temperature of a gas. This exact congruence between the equations justifies one's *identifying* pressure with the force exerted by the tiny particles on the walls of their container.

4. *The Type of Reduction Likely in Psychology*

We are now in a position to ask which one of the above accounts is an accurate description of the relation between

[35] See R. Zaffron, 'Identity, subsumption, and scientific explanation', *Journal of Philosophy,* LXVIII, 23 (2 Dec. 1971), 849–60.

[36] See Enç, 'Identity statements and micro-reductions'. On such formal isomorphism see Carl G. Hempel, 'Aspects of scientific explanation', in his *Aspects of Scientific Explanation and Other Essays in the Philosophy of Science* (New York: The Free Press, 1965), pp. 435–6.

psychological and physiological explanations. Although the example of reactive depression strongly suggested that reduction will be a matter of invalidating and abandoning psychological explanations, it is now clear that not all instances of explanatory reduction are of this type. How does one come to an assessment of the type of reduction likely to occur between psychology and the brain sciences? One needs to consider psychological models which have been related to neurophysiological mechanisms, and to consider the precise form of that relation. The models must be well developed in the sense that the psychological terms must have relatively determinate truth conditions, and they must be relatively successful in accounting for the data.

I believe that some current psychological models of motivation and learning in animals meet these criteria, and I shall use such models as a starting point of analysis.

It may be thought that such a move abandons the problem posed by our example of depression. What do such animal models have to do with human consciousness? I think that only by considering them can one test some of the assumptions and arguments concerning depression. The difficulty with using reactive depression as an example of the logic of reduction is simply that reactive depression is an extremely complicated psychological phenomenon. Many of its psychological aspects are not understood. Depression involves attitudes, beliefs, desires, values, and moods, and those phenomena in turn involve very sophisticated human linguistic and cognitive capacities. A physiological model of depression can account for only those features of depression which are understood psychologically, and presently that restricts such models to some relatively simple learning phenomena. This is why physiological models of depression do not even attempt to touch the complexities of mood, feelings, beliefs, self-perceptions, or values involved in depression.

Perhaps it is this absence of a sophisticated physiological account which in part leads us to believe that psychological and physiological accounts are incompatible. The gap in the physiological explanation may be due more to the incompleteness of our knowledge than to a basic incompatibility of conceptual schemes. Perhaps the physiological story could come to include the phenomena of moods, feelings, and beliefs, once an adequate psychological theory of those capacities is found.

Models of motivation and learning in animals deal only with some of the simpler features of depression, but they have the virtue of being well developed and of having a neurophysiological interpretation. Such models provide a simple analogue for the relations of psychological accounts of human depression to physiology. If our understanding of the logic of reduction is oversimplified even for theories of rat psychology, it is unlikely to be accurate when projected to future models of human psychological capacities.

More strongly, I shall use the simpler models as counterexamples to various pervasive philosophical assumptions about explanatory reduction and the replacement and incompatibility of explanations. I hope to show that some of these assumptions do not even apply to the relations between rodent psychology and rodent physiology. Discarding them will hopefully remove some of the confusions from arguments about reductionism in psychology. The simpler examples are far complicated enough, and an understanding of them will give us a much more sophisticated view of the meanings and consequences of reductionism in psychology.

II

Black Box Models and Neural Mechanisms

A classical requirement for inter-theoretic reduction is that the explanations of one of the disciplines be related to the explanations of the other; to do this requires *connecting principles* relating terms in the two disciplines. An analysis of the way psychological terms are related to terms in neurophysiology would I think resolve many of the problems about explanatory reduction. It would, for example, alleviate the discomfort caused us by the question 'How can the firing of neurons be *painful*?'.

The predominant kind of statement which physiological psychologists use to relate psychological to neurophysiological terms are statements of *localization of psychological function,* and I believe that a sure grasp of the logic of functional localization will give us the nutcracker we need to crack the nut of explanatory reduction in psychology. *Psychological functions* are named 'reinforcement', 'arousal', 'inhibition', 'anticipation', and so on; a *statement of functional localization* claims that such-and-such a neural structure 'fulfils', 'mediates', 'subserves', 'underlies', or 'is the neural basis of' the given psychological function. For example:

> The reticular formation is the neural basis of arousal and cortical desynchronization.
> The medial forebrain bundle mediates natural rewards.
> The orbito-frontal cortex subserves functions of unlearning.
> The ventromedial hypothalamus is the satiety centre.
> The hippocampus is the underlying basis of habituation.

Currently this is the only sort of statement which can act as a connecting principle relating psychological terms to neurophysiological terms, and I suggest that an analysis of this kind of statement will clarify many of the problems of the logic of reduction in psychology. To do this, we must first understand what psychologists mean by the word 'function', and that requires in turn that we understand the form of psychological models.

In the remainder of this book I shall consider primarily psychological models of motivation and learning in animals, and any claims I make about psychological discourse or the meaning of psychological terms are meant to apply only to the discourse within this restricted area of investigation. Psychological models in this area are used fairly strictly; the concepts of sufficient controls, behavioural predictions, and deciding experiments between models can be applied to them. Usually this style of theorizing is not so strict as to require formalization in a deductive calculus or mathematization, yet experimental work is taken sometimes to provide sufficient controls and to decide between rival models. It is for this reason that models of animal motivation and learning were chosen as the starting point of analysis; models in different fields may or may not have similar features.

1. *Black Box Models*

Experimental psychologists investigating motivation and learning try to explain what I shall call 'stimulus-response relationships'. Experimental findings in such studies consist largely of empirical relationships found to hold between characteristics of stimuli impinging on an animal's sensory surfaces and characteristics of its responses. Many of the experimental findings and observations in this field are roughly of the form: in such-and-such stimulus situations the animals will do so-and-so. The stimulus characteristics mentioned may include long sequences of conditioning trials, food deprivation, and so on; and the response characteristics are usually some measurable or countable features of responses.

Here are some examples of stimulus-response relationships:

Rats deprived of food will cross electrified grids in order to obtain food, and the longer they have been deprived of food, the higher the voltage they will endure.

Food-deprived rats which are rewarded with a food pellet for pressing a bar will learn to press the bar.

Food-satiated rats allowed to explore a maze in which some food is located in the goal box, and then deprived of food and placed in the start box of the maze, will run to the goal box significantly faster than food-deprived rats not given the earlier opportunity to explore.

Stimulus characteristics in such findings are specified in terms of experimental procedures which the animals have undergone; the problem of inferring from these procedures to events at the animal's sensory surfaces will be explored below.

Much of the terminology in learning and motivation experiments refers to conditioning procedures which animals undergo, or to arrangements of contingencies in which their behaviour is tested. For example, in 'partial reinforcement' a food-deprived animal is sometimes, but not always, rewarded for pressing a bar by being presented with a food pellet. In 'avoidance learning' tasks an animal must perform a particular kind of response in order to avoid an aversive stimulus such as shock, while in 'escape learning' the animal cannot avoid the shock, but must perform a particular response in order to terminate it. Terms such as 'Sidman avoidance', 'two-way active avoidance', 'reversal tasks', and 'discrimination tasks' can be defined in terms of procedures and equipment set up by the experimenter. Such terminology allows us to abbreviate descriptions of some stimulus-response relationships:

Partially-reinforced rats extinguish more slowly than rats on continuous reinforcement.

Associating a tone with a shock off the baseline leads to an increase in Sidman avoidance rates when the tone alone is presented on the baseline.

Sodium amytal during conditioning abolishes the partial reinforcement extinction effect.

Special terminology is also used to describe features of responses or of stimuli. For example, a 'positive reinforcer' is any stimulus whose presentation increases the probability of recurrence of some immediately-preceding response. A 'negative reinforcer' or 'punishment' decreases that probability. Features of responses in experimental situations are also given names, such as 'conditioned suppression' (rates of bar pressing for food on the baseline drop when a conditioned stimulus for punishment is presented), 'habituation', 'frustration effect', and 'counter conditioning'.

Like any science, psychology must start from observables; and in an investigation of animal behaviour the observables comprise the stimulus situations of the animals, the experimental manipulations they undergo, and certain observable or measureable chacteristics of their responses. To begin, the science requires a stock of concepts whose meanings are fixed and whose application finds reliable inter-subjective consensus. Everyone agrees (more or less) about the observations.

There is a problem, however, in considering the observations to be of stimulus-response relations. Although one can describe in detail the conditioning trials of animals, the controls one sets up, and the test procedures one has arranged, in doing so one is merely recording those aspects of the animal's history which are observable by humans. To talk of stimuli impinging on the animal's sensory surfaces requires certain inferences. One may not have conclusive grounds for believing that the conditioning difference between the experimental group and the control group is the only important difference between the relevant stimuli impinging on the receptor surfaces of the two groups. Typically one hopes that the only important difference between the experimental and the control group will be a difference in the conditioning trials the two groups undergo, so that some controllable difference in the stimulus histories of the two groups can be related to the responses of each group. But from the observable differences in experimental procedures one cannot infer directly to differences in stimulus histories; significant uncontrolled stimuli may be contaminating the results. For example, smell signals left by rats must be carefully considered in some learning experiments; yet sometimes they are not (perhaps because experimenters are primates and do not have the olfactory sensitivities of rodents!).

Well-attested stimulus-response relationships can sometimes

be formalized in equations relating variables in the stimulus history to variables in the response of the animal. The learning curves found in almost any experiment on learning are a typical example. Given the experimental procedures carried out on experimental and control groups, response over time (or some other variable, such as reinforcement intensity or strength of shock) for the experimental group follows one curve, while that for the control group follows another.

A simple example of a class of stimulus-response relationships requiring some explanatory model are those involved in habituation. An 'orienting response' is defined as any modality-non-specific response to novel stimuli which disappears after repeated presentations of the stimulus; it includes such changes as increased heart rate, drop in galvanic skin resistance, increased respiration, cortical desynchronization, and orientation and direction of attention towards the stimulus. The disappearance of this orienting response after repeated presentation of the stimulus is called 'habituation'. Some of the stimulus-response relationships in need of explanation are as follows:

1. If, after an animal has habituated to a stimulus, it does not receive the stimulus for a long period, then when the stimulus is again presented the orienting response spontaneously reappears.
2. If an animal has habituated to a stimulus and the stimulus is presented with some second stimulus, then the orienting response to the former stimulus returns. Habituation to it is re-established more quickly than its original habituation.
3. If the intensity or duration of a stimulus to which an animal has habituated is decreased, the orienting response returns.
4. If an animal has habituated to a compound stimulus (a tone and a light presented together, for example), then presentation of one of the stimuli by itself will restore the orienting response.
5. Change in the frequency or amplitude or other characteristics of a stimulus to which an animal has habituated will restore the orienting reflex.
6. If an animal becomes drowsy during the course of

presentation of a stimulus to which it has habituated, the orienting response returns and is more intense than it was originally.

7. Complete omission of a stimulus to which an animal has habituated when previously presented at regular intervals produces an orienting response at the time when the stimulus would have occurred.

To account for these various stimulus-response relationships, two different models, those of Sokolov and Horn, have been proposed. They describe mechanisms whose input-output relations purportedly would manifest the above seven features.[1] (See Figure 2.7 on p. 60.)

To explain their experimental findings, psychologists think of the organism as a 'black box' whose input-output relationships (at least within some carefully-defined experimental situations) are known, but whose internal structure must be inferred. The task of explanation lies in deciding what sort of machinery inside the black box could produce the responses in question, given the inputs. Ideally, given sufficient knowledge of that machinery, behaviour could be predicted as a function of inputs and internal states of the system. This internal machinery can be specified in different ways, some of which I shall analyse.

Symbolically, a black box is a box with an 'input' arrow entering it and an 'output' arrow exiting from it. The input comprises the variables in the organism's stimulus history which are important for the behavioural phenomenon in question, the output the measured or observed characteristics of its response. The box itself represents a mathematical function or transformation relating the input variables to the output variables.

A structure or mechanism inside the black box is proposed to account for the transformations relating input variables to output variables. One proposes hypothetical states inside the black box--internal variables, whose variation accounts for the observed regularities. These internal variables are themselves related to the input variables and the output variables by a collection of mathematical transformations. Certain parame-

[1] See J. A. Gray, *Elements of a Two-Process Theory of Learning* (London: Academic Press, 1975), pp. 5-10.

ters relate some of the internal variables to the observed input variables, and some of them to the observed output variables. One also hypothesizes parameters relating some of the internal states to other internal states. Some of the hypothesized internal variables are related to observable inputs or outputs only by a sequence of other hypothesized internal variables and hypothesized parameters; that is, they are relatively remote from both stimulus and response variables, and are related directly only to other internal variables. (See Figure 2.1.)

If one thinks of the model as a mathematical structure, therefore, in proposing a mechanism accounting for the observed regularities, one is analysing the equation relating input to output into many internal equations relating hypothesized internal states to one another and to the input and output variables. A mathematical function (which I shall symbolize by 'function$_m$') is a mapping from one set to another which is not a one-many mapping. If the output is a function of the input, then one analyses that function$_m$ in terms of hypothesized internal states related by other functions$_m$ to the inputs, the outputs, and each other. The operation of the intervening machinery produces the function$_m$ relating observed inputs to observed outputs.

The primary domain of mathematical models in psychology is in psychophysics, although one finds some mathematical models in motivation and learning theory. The Gray-Smith model of classical and instrumental conditioning was given a mathematical formulation.[2] Toates and Oatley have proposed a mathematical model for thirst which has been instantiated in an analogue computer simulation, whose behaviour fits some of the experimental data quite well.[3]

Another way of specifying the black box machinery is with a machine table. A machine is an entity whose output is a function$_m$ of its input and its internal states; a machine table represents the output state of the machine for every combination of input states and internal states.[4] The machine table is

[2] Ibid., pp. 355–83.

[3] D. J. MacFarland, *Feedback Mechanisms in Animal Behaviour* (London: Academic Press, 1971), pp. 169–79.

[4] On machine tables see Hilary Putnam, 'Minds and machines', in his *Mind, Language and Reality*, Philosophical Papers, Vol. II (Cambridge: Cambridge University Press, 1975).

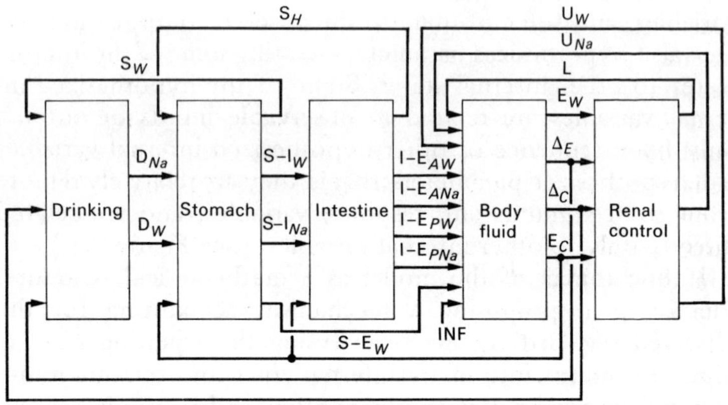

A. Thirst sub-systems and their interconnections

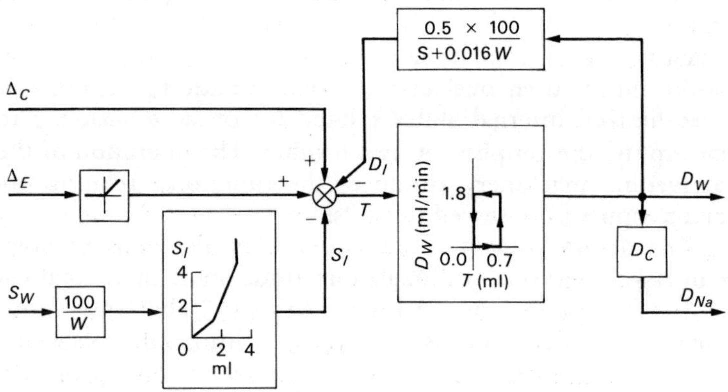

B. Drinking sub-system

- D_C Drinking fluid concentration of sodium (m-equiv./ml)
- D_I Drinking inhibition factor (ml/100 g body wt)
- D_{Na} Rate of drinking sodium (m-equiv./min)
- D_W Rate of drinking water (ml/min)
- Δ_C Cellular error (ml/100 g body wt)
- Δ_E Extracellular error (ml/100 g body wt)
- S_I Stomach inhibition factor (ml/100 g body wt)
- S_W Stomach water (ml)
- T Total thirst signal (ml/100 g body wt)

Fig. 2.1. Part of a mathematical model of drinking. Each sub-system relates its input variables to its output by various mathematical functions. Only the variables and the internal parameters of the drinking sub-system are shown. (From F. M. Toates and K. Oatley, 'Computer simulation of thirst and water balance', *Medical and Biological Engineering*, 8 (1970), 73, 81. Reprinted by permission of Pergamon Press Ltd.)

an abstract characterization of the mechanisms within the black box which are such that, when it is in the particular internal state hypothesized, then given those inputs, the appropriate output will be produced.

A black box approach asks the question 'What sort of machinery must there be inside the black box to produce that behaviour, given that stimulus history?'. By 'machinery' what is meant is some internal structures which are related to each other in such a way that, given the internal states and the stimulus history, the observed responses are causally produced. The internal parts are specified by the sorts of jobs they do relating inputs to outputs. Internal states within the mechanism are interrelated by certain transformations, and related to input and output states; these relations between the states are produced by the workings of the hypothesized internal parts.

A third way of specifying the machinery inside the black box is with what I shall call a *black box diagram.* Such a diagram consists of a number of boxes connected by arrows. The lines connecting the boxes represent the hypothesized internal variables. The boxes represent transformations relating these variables to each other. With each box is associated a function$_m$ relating its inputs to its output; each box maps its input to its output and sends that output to other boxes.

Consideration of an example of a black box diagram may clarify these comments (see Figure 2.2). In this diagram, $s_1 \ldots s_4$ are stimulus variables and $r_1 \ldots r_4$ are response variables, while the boxes and lines between them represent a model invoked to explain regularities they display. Variables $x_1 \ldots x_7$ are the internal variables in the system, and each box represents a certain mathematical function relating its inputs to its output. For example:

$$x_1 = f_1(s_1, s_2)$$
$$x_5 = f_5(x_1, x_2)$$
$$r_3 = f_{10}(x_6, x_7)$$

and so on. Since each box represents a mathematical function relating its inputs to its output, each can have only one output variable; if there are two distinct output variables then we know that there are two functions$_m$ relating those outputs to the input of the box, and hence, two boxes. By the *antecedent variables* of a given variable I shall mean all of the inputs into the box of which the given variable is the output. For example, the antecedent variables of x_5 are x_1 and x_2. By the *consequent variables* of a given variable I shall mean all the outputs of the boxes for which the given variable is an input variable. So x_5 and x_7 are consequent variables of x_2. These are the immediate variables which changes in x_2 will influence, while the antecedent variables of a given variable are those which determine its value.

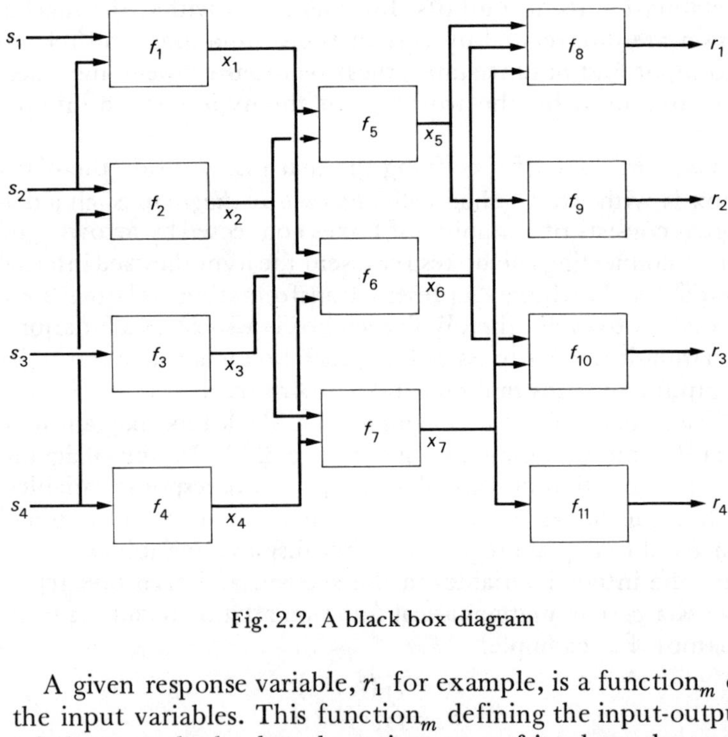

Fig. 2.2. A black box diagram

A given response variable, r_1 for example, is a function$_m$ of the input variables. This function$_m$ defining the input-output relations can be broken down in terms of its dependence on variations in internal variables. For example, r_1 has as antecedents x_5, x_6, and x_7; x_6 has as antecedents x_1, x_3, and x_4, while x_7 has as antecedents x_2 and x_4; and these antecedents of x_6 and x_7 in turn depend on the stimulus variables.

One point that must be emphasized is that in saying that

the variables in the models are states of the pathways relating boxes in the model, one is not committed to the claim that all the variability in the model arises from variation in *neural* pathways. At this stage, the model presents a merely functional characterization of the system, and its precise relation to the nervous system is left indeterminate. The goal is to relate both the boxes and the pathways in the model to collections of neurons, but one must not assume that the pathways in the model necessarily correspond to anatomically identified neural pathways such as the medial forebrain bundle or the amygdalofugal pathway. The neurons corresponding to a pathway in the model may not be those of any such anatomically-identified pathway. (See below, pp. 101-4.)

Similarly, the claim that the boxes in the model map their inputs on to an output by some constant function does not commit one to the claim that there is no variability in the functioning of neural structures such as the hippocampus or septum. The boxes in the model cannot be presumed to identify such anatomically-identified structures; their neural identity is at this point unspecified. The fact that the functioning of a neural structure can change therefore should not be taken to defeat a model. These kinds of variability in the system can themselves be captured in black box analysis.

Any hypothesized internal state or variable in the system is related in certain ways to other internal states and to input and output variables. In terms of machine tables, we could specify the necessary and sufficient conditions for occurrence of a given internal state by specifying a disjunction of all the combinations of inputs and internal states sufficient to prompt the occurrence of that state. The consequences on behaviour of each of these combinations could then be inferred. Except in very simple models of neural networks, theorists rarely actually write a machine table; instead they specify a variety of internal states thought important in the animal's behaviour, and then (ideally) specify the necessary and sufficient conditions and the effects of the occurrence of each. These conditions and effects are specified in terms of stimulus conditions, response variables, and other hypothesized internal states. (See Figure 2.3.)

The disadvantage of this approach is that it assumes the

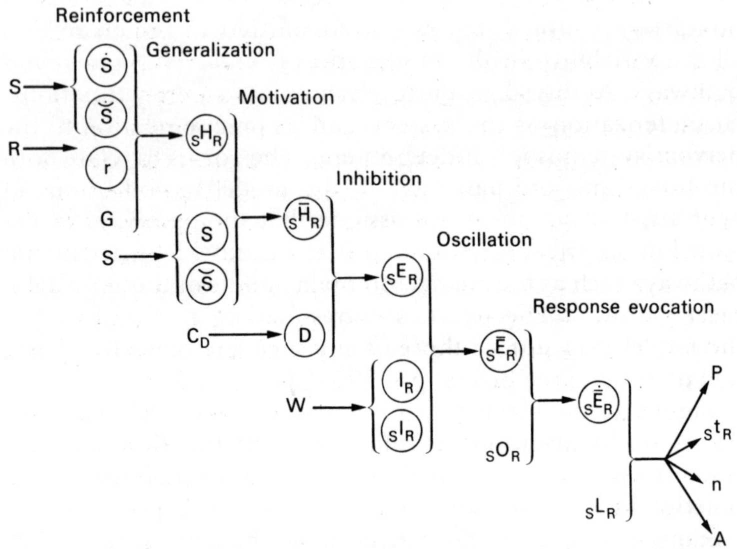

- S — The physical stimulus energy
- R — The organism's reaction
- $\dot{S}$ — The neural result of the stimulus
- $\breve{\dot{S}}$ — The neural interaction arising from two or more stimulus components
- r — The efferent impulse leading to reaction
- G — Occurrence of reinforcing state of affairs
- $_SH_R$ — Habit strength
- S — Evocation stimulus on same stimulus continuum as S
- S — The neural result of S
- $\breve{S}$ — Neural interaction
- $_S\bar{H}_R$ — The generalized habit strength
- C_D — The objectively observable phenomena determining the drive
- D — The physiological strength of the drive
- $_SE_R$ — The reaction potential
- W — Work involved in evoked reaction
- I_R — Reactive inhibition
- $_SI_R$ — Conditioned inhibition
- $_S\bar{E}_R$ — Effective reaction potential
- $_SO_R$ — Oscillation
- $_S\dot{\bar{E}}_R$ — Momentary effective reaction potential
- $_SL_R$ — Reaction threshold
- P — Probability of response evocation
- $_St_R$ — Latency of reaction evocation
- n — Number of unreinforced reactions to extinction
- A — Amplitude of reaction

Fig. 2.3. A psychological machine table: Hull's 1943 model. Although this is not a complete machine table—because it does not specify every combination of values of all the variables—nevertheless it does present the antecedent and consequent variables for every internal variable in the system. (From Clark L. Hull, *Principles of Behavior* (New York: D. Appleton-Century, 1943), p. 383. Reprinted by permission of Richard H. Hull.)

internal variables are not continuous; otherwise, one can more easily state the relationships between them in terms of mathematical equations. The black box diagram also represents a machine table, in that it represents the conditions for the occurrence of any given variable in the system and the conditions which result when it occurs. The advantage of both the black box diagram and the mathematical system is that the machine table must list each distinct value of each variable, and so is extraordinarily cumbersome. In effect the machine table says 'variable *x* will attain value *g* if and only if . . .' while the mathematical structure and the diagram do not need to specify discrete values of the variables. The important point is just that one can use either a system of equations, a machine table, or a diagram to specify the mechanisms in the black box.

Different models propose different organizations of internal states–different machine tables–to account for the given stimulus-response relations. For example, on the Miller-Mowrer model of avoidance learning, the decrease in fear which has been conditioned to the start box reinforces the animal for making the response. In Gray's model, avoidance behaviour is explained by the positive reinforcements of safety signals—conditioned signals of non-punishment which the animal receives after making the response. On this view the non-occurrence of a punishment is a positive reinforcer in its own right. The two models describe differently the internal states involved in avoidance behaviour; the internal states thought to be involved have different antecedents and different effects on response variables, although their combinations purportedly account for the same stimulus-response relations.[5] (See Figure 2.4.)

Each internal variable in the model has certain relations to some antededent and consequent variables. This is unsurprising, since such internal variables are invoked to explain stimulus-response relations; it is alterations in the internal variables which account for changes in behaviour. The stimulus-response relationships characteristic of the system sum all the relations between the internal variables, and between the internal variables and the stimulus and response variables.

Some of the terms in psychological models of learning and

[5] See J. A. Gray, *The Psychology of Fear and Stress* (London: Weidenfeld & Nicolson, 1971), Chs. 11 and 12.

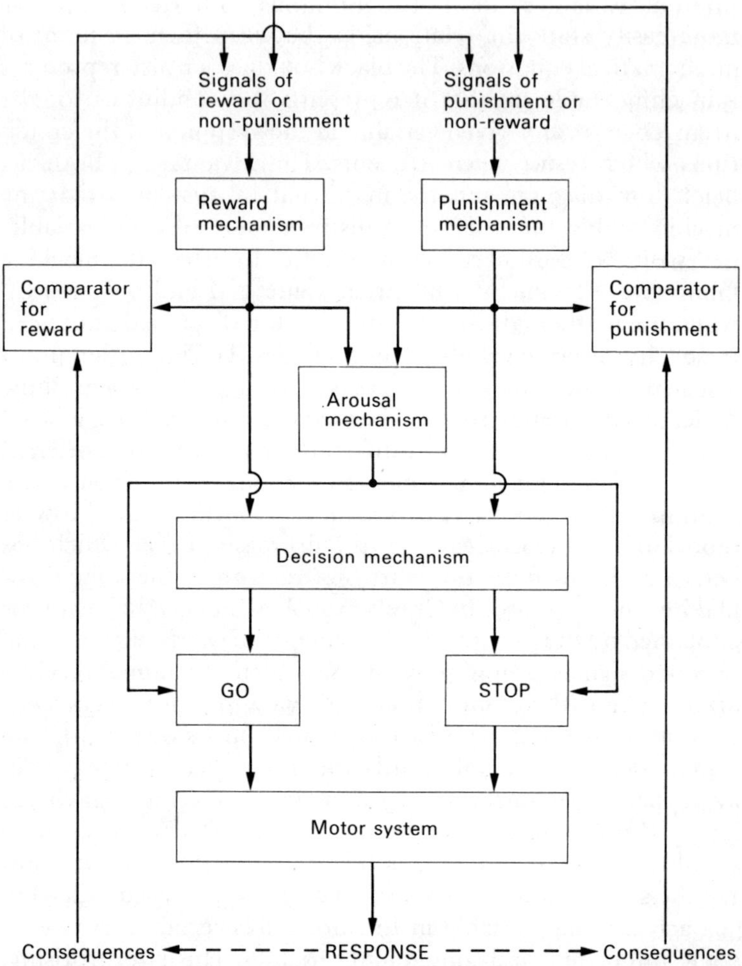

Fig. 2.4. Gray's two-process model. The comparators can detect non-punishment or non-reward only after expectancies have been formed by classically conditioning rewards or punishments to initially neutral conditioned stimuli. Rewards or punishments select behaviour (through the decision mechanism) by instrumental conditioning. From J. A. Gray, *Elements of a Two-Process Theory of Learning* (London: Academic Press, 1975), p. 347. Copyright by Academic Press Inc. (London) Ltd. and reprinted with their permission.)

motivation name internal variables. For example, 'hunger' or 'thirst' presumably can range from zero value up to some physiologically determined maximum. Hours of food deprivation, rate of metabolism, amount of muscular activity, amount of fat storage, and so on, are antecedent variables for 'hunger'. Changes in hunger state alter such consequent variables as: degree of arousal, degree of reinforcing value of food, degree of goal-directed activity towards food, and so on. (The specific antecedent and consequent variables for a term such as 'hunger' or 'drive' can only be given by a specific model.) Another example is the term 'arousal', whose level is determined by antecedent variables such as: drive level, stimulus intensity, novelty of stimuli, and neural factors such as fatigue. Arousal in turn has consequences on such variables as: general alertness, intensity of response, efficiency of learning, and efficiency of response.[6] Expectancies, secondary reinforcers, conditioned drives, fear, safety signals, and so on, are other states invoked in models of motivation and learning; they correspond to values of hypothesized internal variables in the black box diagram, and each is related to certain other variables upon which it depends and others which it affects.

A caveat to this account of variables in a model must be noted. Some of the terms used to name internal variables in a black box model are sometimes used instead to name the activity of the box whose output is the given variable. For examples, 'arousal' is sometimes used to refer to a process occurring inside the black box; in this use it names the activity of the part whose output is the internal variable 'arousal level'. The terms are used to name psychological *functions,* which here means: the activity of the parts responsible for imposing the transformations on inputs to produce the variable in question. 'Reward' or 'drive' are other examples of terms with this twofold use.

These first three characterizations of psychological models—as systems of equations, machine tables, or black box diagrams—apply only to some of the complicated species of the genus. We can find a simpler way of talking about such models by considered again the starting point of black box analysis. Psychologists explain given stimulus-response relations by

[6] Gray, *Two-Process Theory,* p. 93.

postulating, inside the black box, a mechanism whose structures and workings are interrelated in such a way that the system as a whole will manifest the appropriate input-output relationships. They ask 'What sort of machinery is required to produce these observed relations? What sorts of jobs must parts inside the black box fulfil for it to manifest these outputs?'. To answer these questions the psychologist postulates hypothetical machine parts, arranged in a particular way, which work together to produce the observable regularities.

An example of a mechanism proposed to account for some stimulus-response relationships is one proposed to account for feeding behaviour. There are many stimulus-response relationships which such a model must explain, but two are central. The first is that over the long term, rats and other animals maintain themselves in a state of *energy balance*. That is, the net amount of energy they absorb from radiant and thermal energy and the chemical energy of their food is equal to the amount of energy they expend in metabolic and muscular activities. That this is so is evident from the fact that the rat neither gains nor loses body weight on the long term. In fact, not only is energy balance maintained, but so are a variety of other homoeostatic conditions: the rat sustains a fairly constant ratio between kinds of protein it ingests, and a constant level of vitamins and minerals. Given a choice of diets, rats will eat a balanced one: sodium-deprived rats will choose diets rich in sodium, and rats will learn to choose diets rich in calcium, potassium, or thiamine if they are deprived of those substances.[7]

A second sort of behaviour a model of feeding must explain is that at times rats and other animals *seek food;* they engage in goal-directed behaviour aimed at the ingesting of foodstuffs. Their behaviour is goal directed in the sense that they will compensate for environmental variations and adapt their behaviour in order to reach the same end-state. In order to obtain food they will learn to perform instrumental responses. When hungry, rats will show a primitive level of insight or reasoning in maze problems, in order to obtain food most efficiently.

[7] Peter M. Milner, *Physiological Psychology* (London: Holt, Rinehart & Winston, 1970), pp. 317–21.

To some philosophers this seems to pose a problem for any mechanical account of behaviour. When we explain the maze-learning behaviour of the rat by saying that it is 'seeking food', we are offering a telelogical explanation of the rat's behaviour, claiming that the rat has a particular goal (ingesting food) and that it performs that behaviour *in order to* achieve that goal. The behaviour in the maze is a means towards the end of obtaining food. This seems to involve the claim that some representation of the goal guides the rat's choices; because it is seeking food, the rat will act in certain ways and not in others. (By 'representation' I do not mean a symbolic representation; it may be simply a neural phenomenon.) We seem to be explaining the behaviour of the rat by attributing to it goals which influence its choices: the rat is *seeking food,* and this goal influences its behaviour.

However, these difficulties are not real, at least for this sort of simple goal-directed behaviour; both sorts of behavioural phenomena are accommodated by the elegant hypothesis that a negative feedback mechanism controls feeding. A 'negative feedback' system is one in which the difference between some optimal state of the system and its actual state controls processes whose effect is to eliminate that difference, and hence restore the system to that optimal state. Nagel presents a formalization of this idea;[8] there is some *G* state such that variations out of the *G* state caused by disturbance variables result in compensatory variations in the system whose effect is to restore the *G* state. In general, any system which maintains itself within a *G* state by compensatory variations is a negative feedback system. We need suppose only that *energy balance* is the *G* state for rats, and that deviations from this optimal state actuate behavioural mechanisms whose effect is to restore the *G* state (by food ingestion). (See Figure 2.5.)

Any homoeostatic drive (for example, hunger, thirst, or temperature regulation) involves some *G* state; signals of deviation from the *G* state initiate behaviour whose effect is to restore the homoeostasis. Homoeostatic drives involve negative feedback mechanisms, which account both for the maintenance of the given *G* state and for the goal-directed behaviour restoring them. The central physiological question is:

[8] Nagel, *The Structure of Science,* pp. 411–18.

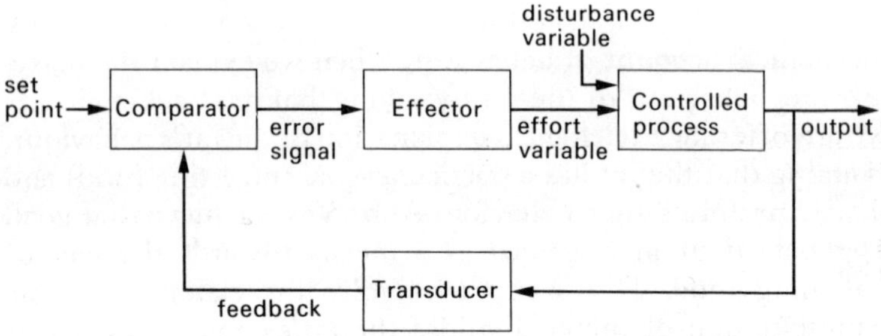

A. General design of a negative feedback system

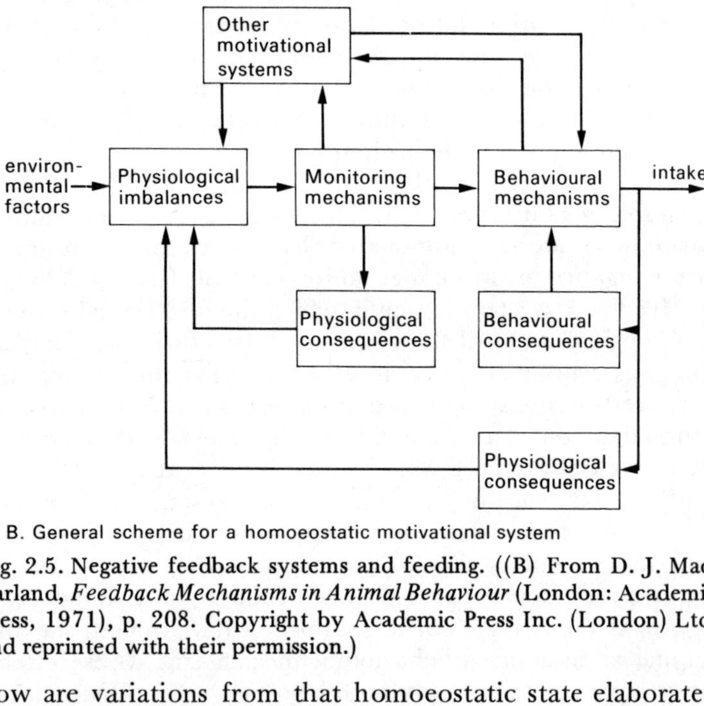

B. General scheme for a homoeostatic motivational system

Fig. 2.5. Negative feedback systems and feeding. ((B) From D. J. MacFarland, *Feedback Mechanisms in Animal Behaviour* (London: Academic Press, 1971), p. 208. Copyright by Academic Press Inc. (London) Ltd. and reprinted with their permission.)

how are variations from that homoeostatic state elaborated into behaviours which restore the state?

This hypothesized mechanism embraces some of the psychological states proposed in models of feeding. Drive states, for example, may be error signals—signals of deviation from optimal energy balance—which set the behavioural mechanisms into action. Drive reaches zero level when the animal attains satiety. We would expect reinforcement to be involved whenever a behaviour results in a decrease in the deviation from the

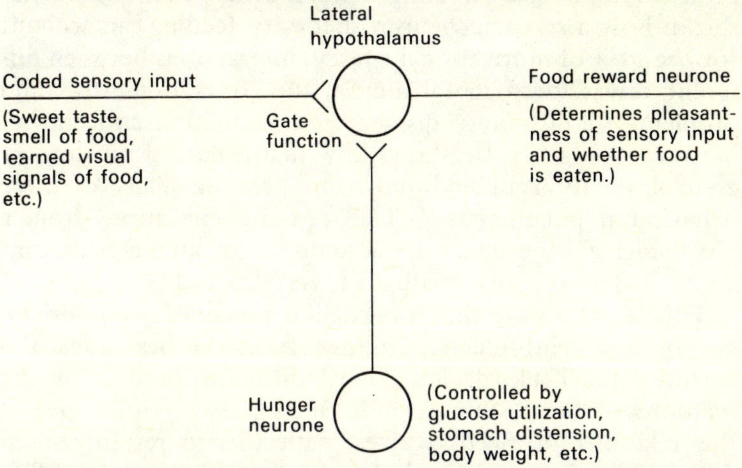

Fig. 2.6. Rolls's neural gate model of drive and reward. 'The effects of coded sensory input on lateral hypothalamic reward neurones are gated by neurones which signal hunger. The input is coded beyond the receptor level so that, for example, the smell of food but not of non-food objects is signalled, or, for example, the learned sight of a food but not of a non-food object is signalled. The gating (which could be pre- or post-synaptic) by "hunger" neurones is perfomed on the basis of factors such as glucose utilization, stomach distension, and body weight. Activity in the food reward neurones determines the pleasantness of the sensory input and whether feeding will continue.' (From E. T. Rolls, *The Brain and Reward* (Oxford: Pergamon, 1975), p. 28. Reprinted by permission of Pergamon Press Ltd.)

G state—that is, whenever need reduction occurs. In fact this is not always so; however in most models there is some relation between needs, reinforcement, and satiety. For example, Deutsch's model relates drive and reinforcement as follows: a reinforcement is any stimulation of an analyser whose link has been set into activity by drive excitation from the primary link.[9] (See Figure 4.1 on p. 135.) In Rolls's model, drive stimuli act upon a gate system which controls the degree to which peripheral stimulation excites reward neurons in the hypothalamus.[10] (See Figure 2.6.) Instrumental learning occurs, on the negative feedback model, because of the

[9] J. A. Deutsch, *The Structural Basis of Behaviour* (Cambridge: Cambridge University Press, 1964), p. 54.

[10] Rolls, *The Brain and Reward*, p. 28.

presence of a goal-directing drive. Finally, MacFarland has shown how a servomechanism theory of feeding can account for the idea of motivational energy, interactions between different drives, behavioural inertia, the frustration-effect and problems of dithering, displacement activities, and choices between two goals. He shows how mathematical portions of control theory can be brought to bear on many of these behavioural phenomena.[11] The idea that deviations from a physiologically optimal state actuate behavioural mechanisms controlled by negative feedback is very fruitful.

This is not to say that a particular model of needs, drives, satiety, and reinforcement follows from the basic idea of a negative feedback device. Many different models of the relations of these states have been proposed. Hull's idea of the relations of needs, drives, satiety, and reinforcement differs from Tolman's; and Morgan, Bindra, Deutsch, Miller, Rolls, and Gray have all proposed different accounts of these states.[12] Each of these proposals is, however, congruent with the idea of a negative feedback device controlling feeding.

The important point is that the teleological explanations of the rat's behaviour are perfectly compatible with a causal account; the foodseeking behaviour of the rat is not in principle any more mysterious than the heatseeking behaviour of anti-aircraft missiles. In both, the goal-directedness can be explained by a negative feedback mechanism which controls behaviour depending on the size of the difference between its actual state and its *G* state. In both, some representation of the goal influences behaviour, in the sense that some part of the mechanism (a comparator) computes the difference between optimal and current states, and sends error signals which control behaviour.

The hypothesized mechanisms offered by such models are rarely given a detailed structural characterization, but are instead described in terms of hypothesized parts having certain jobs, related in certain ways. Typically the description of a part proceeds entirely by describing its relations to other parts and the contribution it makes to the workings of the system as a whole.

[11] MacFarland, *Feedback Mechanisms in Animal Behaviour*, sections 7.2, 7.3.

[12] See Fred P. Valle, *Motivation: Theories and Issues* (Monterey, California: Brooks/Cole Publishing Company, 1975), for an analysis of these different models.

For example, one model for habituation supposes that within the black box, part of the mechanism builds up a neuronal representation of the stimulus, and another part compares this representation to the incoming stimulus. A match inhibits the parts responsible for organizing the orienting response, while a mismatch sets them into action. Another model of the phenomenon suggests that each stimulus sends an impulse down a pathway to the mechanism which initiates the orienting response, but if the same stimulus is repeated a number of times, the strength of the signal reaching that mechanism gradually decays, and so habituation is produced. Changes in stimulus parameters change the signal features and offset the decay in strength of the signal.[13] (See Figure 2.7.)

What is important in this example is that the hypothesized parts of the mechanism are specified in terms of the jobs they perform: one part builds up a neuronal representation, another makes a comparison, and another initiates the orienting reflex. Usually the boxes in a black box model represent hypothetical parts of the system specified by their functional role—by the jobs they do, the consequences which their activities have on the system as a whole. Such parts are given structural relations to other parts only to the extent that is necessary for an understanding of the way the parts work together.

The characterization of the hypothetical mechanism is abstract in the sense that one can imagine different sorts of structural arrangements which fulfil all the required functional roles. The machine may be made of nerve cells or of transistors and solid-state circuitry, and yet produce identical input-output relations. For a given box such as 'mechanism for neuronal representation', one can invent different arrangements of neurons which will do the job.

'Psychological function' often means: a job performed by a hypothesized part of the mechanism inside the black box. This will be clarified in the next section. But first a final feature of black box models in motivation and learning must be considered.

Along with its theoretical overtones, the question of 'What goes on inside the black box?' has a literal sense: the inputs to the black box are physical events at the receptors, the

[13] Gray, *Two-Process Theory*, pp. 10–29.

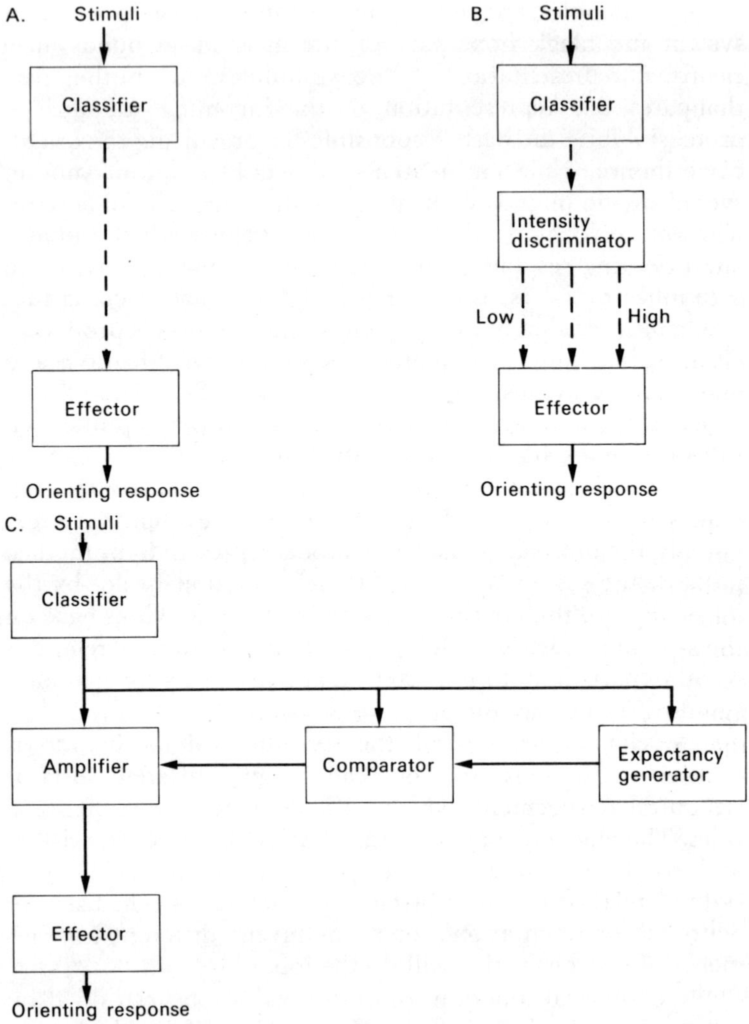

Fig. 2.7. Habituation: Sokolov and Horn (A) Horn's basic model. Habituation occurs along the pathway between the classifier and the effector mechanism. (B) Horn's model for dishabituation due to change in stimulus intensity. Stimuli of different intensities reach the effector mechanism through different pathways, as selected by the intensity discriminator, and habituation occurs in the pathways between the discriminator and the effector. (C) Sokolov's model of habituation. (From Gray, *Two-Process Theory*, pp. 12, 15, 28. Copyright by Academic Press Inc. (London) Ltd. and reprinted with their permission.)

outputs are effector events, and processes in the nervous system spatially and temporally intervene between these two events. Inputs literally 'flow into' the black box, in the sense that processes caused by physical events at the receptors affect processes located more deeply in the nervous system. The black box in psychology literally corresponds to the organism; events inside the black box are events occurring somewhere within the spatial boundaries of the beast. In so far as one's model of processes intervening between stimuli and responses is correct, those processes will be part of the chain of cause-effect links physically connecting stimuli and responses. They occur 'in' the black box in the double sense that they are hypothetical processes invoked to explain observable regularities and that they are processes spatially located within the organism.

The model therefore must adequately specify the causal factors responsible for the production of given stimulus-response relationships; and presumably these causal factors can be located somewhere in the pathways of the nervous system which lead from stimuli to responses. In that case, some physical states and processes in the nervous system correspond to variables in the model, and some structures in the nervous system correspond to boxes in the model, so that jobs assigned to those boxes are jobs performed by parts of the nervous system. This will be explained further in Sections 4 and 5.

2. *Different Meanings of 'Function'*

Some terminological distinctions will prove useful in what follows. We must distinguish between different senses of the word 'function'—a useful word in discussions of psychological models, provided its senses are kept distinct.

The first sense of 'function' which is important for our purposes is that given the word by mathematics, where 'function' means a mapping between two sets which is not a one-many mapping. I have been designating mathematical functions by 'function$_m$'. Four additional senses of the word are sometimes confused:

(v) A class of patterned and recurring processes characteristic of organisms and necessary for their continued life;

so-called 'vital functions' such as digestion or respiration.

(a) The activities, processes, or operations of a given part or organ within a system, that is, the *functioning* of the part.

(c) The inclusive set of consequences of the state or activity of a part for the workings of the whole or for various other parts of the system.

(g) The contribution which the state or processes of a given part make towards the maintenance of a *G* state in the system; its role in maintaining the system in a steady state.

Some examples will clarify these different senses. Functions$_v$ are processes such as digestion, respiration, circulation, and so on. Psychologists use the word 'function' in this sense when they talk about functions$_v$ such as perception, cognition, or memory. These terms denote processes important to the survival of the creature. Another example is provided by the question: are motivational processes a different kind of psychological function$_v$ from emotional processes?

The other meanings of the term are more important than its 'vital function' sense. They are illustrated by the different ways one can talk, in medical discourse, about the function of a body organ such as the liver. One can use 'liver function' in sense (a) to refer to the processes going on within the liver itself--its ordinary activity. Using the (c) sense of the word—referring to the consequences of the liver's work for the body as a whole—the functions$_c$ of the liver are the jobs it performs, the role it plays in the body. By 'job' or 'role' we mean the consequences which the part's activities have on other parts of the system and on the input-output characteristics of the system. One of the functions$_c$ of the liver is to filter out blood poisons; it decreases the concentration of urea and other metabolic waste products in the blood. This particular set of consequences can be attributed only to the workings of the liver, and so it is one of the functions$_c$ of the liver.

In fact, the effects of the ordinary activity of the liver are more precisely delimited, because a steady state exists: the liver maintains the blood within a certain range of concentration levels for various substances. More formally, there is a *G* state (defined by the concentration levels of certain substances in the blood) such that the disturbance variables which cause the blood to become too toxic cause the liver to make compensatory variations (that is, to filter the blood) which restore it to its *G* state. Only because the liver maintains the organism within a certain *G* state can one say that the function_g of the liver is to maintain the purity of the blood.

Obviously every part of a system which has a function in sense (g) also has a function in sense (c): if the part helps to maintain the system in some determined steady state, its activities must affect the activity of other parts and of the system as a whole. The (g) sense is, however, more restricted; only if there is a determinate steady state, maintained by processes in the system, can a part of the system have a function_g.

When physiological psychologists discuss the localization of psychological functions, are they using the word 'function' in sense (v), (a), (c), or (g)?.

Sometimes, although rarely, the localization question means 'Which structures in the nervous system are necessary for the occurrence of some psychological function_v like perception?'. When they talk about the functional organization of the nervous system, physiological psychologists sometimes refer to a division of the structures responsible for the processes of perception, cognition, memory, motivation, and so on. This sort of talk is not very precise, and questions of whether a particular process counts as an instance of one function_v or another may have no determinate answer. For example: is the classically-conditioned hypothetical internal state responsible for a rat's learning active avoidance a motivational phenomenon ('learned drive') or a cognitive phenomenon ('expectancy')? Without a clear distinction between motivational functions_v and cognitive functions_v, there is no way to answer such questions.

The most important meaning of psychological 'function' derives from black box analysis. A black box model proposes hypothetical structures which work together in such a way as to produce the observed stimulus-response regularities; each hypothetical part in the model is assigned a certain job or role, which contributes to the input-output relations of the whole.

In labelling a box 'comparator', 'arousal mechanism', or 'inhibition system', we are assigning jobs to those hypothetical structures. The jobs are defined by the consequences of the activities of those structures on the processes of the rest of the system. The names for such jobs are thus names of psychological functions$_c$. Localization of function concerns the problem of finding an actual structure in the nervous system which does the jobs (has the functions$_c$) assigned to some hypothetical part in the model.

If the black box model contains a feedback loop, then some of its parts are assigned, not only functions$_c$, but also functions$_g$, since there is some *G* state which those parts work together to maintain. The functions$_g$ of the parts in a feedback loop can be given standard names, such as 'error detector', 'actuator', or 'transducer'. These functions$_g$ can be ascribed to parts of the system only if they contribute to maintaining some *G* state.

A box fulfils its function$_c$ by transforming its input variables into an output, given that its output has the specified effects on other boxes. That is, the jobs assigned to hypothetical parts of the mechanism can be analysed in terms of the transformation of inputs into an output and the effects of that output on other parts. Any such transformation will define a relationship between two classes of variables, and if those variables can be assigned numerical values, some function$_m$ will relate the inputs to the output of the box. In this way the function$_c$ assigned to a box corresponds to the function$_m$ by which it produces an output from its inputs. Our meaning, when we give a functional characterization of the mechanism, is ambiguous: we are either specifying the sorts of jobs which its internal parts must fulful, or presenting a set of functions$_m$ relating its inputs and its outputs.

I shall return to the relations between functions$_c$, functions$_m$, and functions$_a$ after explaining more fully the way in which such black box models are related to the nervous system. But first I shall explain a point crucial to the understanding of function$_c$ concepts.

3. *Behavioural Change and Functional Change*

It is difficult to analyse the ways psychologists use experimental evidence to choose between rival black box models. Similarly, it is difficult to understand how they choose between rival explanations of behavioural changes after some physiological intervention. Both of these difficulties are instances of the general problem of analysing the inferences relating changes in values of internal variables in a model to changes in observed behaviour.

One can invent more than one model to explain a given finite collection of stimulus-response relationships. Two or more models may differ over the sorts of components required or their functions, and yet yield identical predictions over some finite list of experimental findings. Sokolov's and Horn's models for habituation are examples, as are Amsel's and Capaldi's models for the partial reinforcement extinction effect, and the Miller-Mowrer and the Gray models for avoidance learning.[14] To decide between rival models, one must find some experimental prediction made by one of the models but denied by the other, or some stimulus-response relationship which one model can explain but the other cannot. Ideally, the difference in the functional organization of the two models will be manifest in a difference in behaviour in some situation. In some cases one experiment will be taken to decide between two models, because the models make contradictory predictions about its outcome. One example is an experiment designed to discover whether rats prefer unavoidable shock associated with a signal warning of its imminent arrival, or shock at the same frequency without a warning signal. The Miller-Mowrer theory predicts the latter, while Gray's model predicts the former.[15]

I said above that a model specifies components of a mechanism, giving every part a job: each has certain inputs, and maps them on to an output. This output contributes in a certain way to the activity of other parts and thereby affects the input-output functions of the system. To investigate the

[14] On Sokolov and Horn see Gray, *Two-Process Theory,* Ch. 1; on Amsel and Capaldi, ibid., pp. 291–6; on Miller-Mowrer, ibid., Ch. 10.

[15] Gray, *The Psychology of Fear and Stress,* pp. 173–4.

effects of a particular variable in a model one must try to discover controls for as many of the other hypothesized internal variables as possible. To discover the influence of a particular part's activity on the input-output relations, we need ideally to hold constant all of the other internal variables which affect the response variables influenced by that part. For example, if we are investigating the effect of different drugs on the reward value of self-stimulation, we need to be sure that the drugs do not alter arousal level or motor co-ordination, as either of these variables can also affect self-stimulation behaviour. If we want to be sure that a particular structure affects the rage thresholds of an animal, we need to be certain that interventions in that structure do not cause pain, as pain can affect behaviour in ways indistinguishable from the effects of a lowered rage threshold. We need to exclude the possibility that changes in other parts of the mechanism can account for the behavioural change, and to do this, we must try to find sufficient controls.

Controls are intended to weed out variation in any of the other factors which might explain the behavioural phenomenon in question. For example, to see if ulcers are caused by conflict, Sawrey, Conger, and Turrell deprived rats of food and water, and required the experimental group to cross an electrified grid to obtain them. This experiment required eight controls, to eliminate the conflict between shock and hunger or thirst, while holding every other combination of food deprivation, and shock duration and intensity constant. If some variable other than conflict caused ulcers in the experimental group, presumably ulcers would also have been produced in one of the control groups.[16] Only if one has grounds for believing that one's controls have been adequate can one ascribe that behavioural change to change in a particular internal variable.

Given that one has grounds for believing that all of the other relevant internal variables are held constant, any alterations in behaviour are ascribed to the given function. Changes in the activity of a given part must in some instances manifest themselves in behavioural changes; specifically, when one can control for the contributions of the other parts. If this were

[16] Ibid., pp. 135–6.

not so then different models–proposing different arrangements of parts and functions–could not be tested by their different predictions.

The implication of this argument is that a model allows one to specify how an internal variable is changed by antecedent variables, and how its changes affect consequent variables. Ideally the system is such that if one has grounds for believing that all of the other variables affecting a given behaviour have been controlled, then in those situations, if the behaviour occurs, then the variable in question has a certain value. In some cases the behavioural evidence is *sufficient* for the ascription of a given internal state. For example, if in a self-stimulation experiment, rates of self-stimulation drop, arousal and motor co-ordination effects are the only secondary variables influencing self-stimulation, and their influence is ruled out, then reward has been attenuated. If the rates of self-stimulation rise, then reward has been increased. Part of the meaning of 'reward' is: if other internal variables (involved in the behaviour) are held constant, then self-stimulation rates are related directly to reward. If this were not part of the meaning of the term, then one could not predict what would happen if the 'reward' variable were increased, and self-stimulation rates would not provide evidence for levels of reward.

Of course we may discover that rates of self-stimulation depend also on some variable *X* other than reward, arousal, and motor co-ordination, and such a discovery would force a revision in many of the current inferences about reward mechanisms in the brain which are based on self-stimulation experiments. We would then change our use of the term 'reward'; where once we used the term to conflate two variables, we would thereafter recognize them as independent, namely, reward and variable *X*.

However, given the current use of the term, if we have controlled for the other variables we think are involved in self-stimulation behaviour, and some intervention decreases rates of self-stimulation, then it is contradictory to claim both that our model of the behaviour is a good one and that the intervention has increased reward values. Either our model of the variables involved is incorrect, or reward has been attenuated. If we could show that reward has increased, even though self-stimulation rates have decreased, then we would show the model to be incorrect, for we would show that other internal variables, which have not been controlled, must be involved. This example shows how experimental evidence can be brought to bear on a model of the variables involved in a behaviour.

Changes in the value of any internal variable in the model must in some situations be detectable in changes in stimulus-response relationships. The need to make different predictions of stimulus-response relationships from different models has already been mentioned as a reason why this must be so. A second reason is that if there are no circumstances under which changes in a given internal variable result in detectable changes in behaviour, then the variable is entirely superfluous for an explanation of stimulus-response relationships. The variable complicates the model needlessly and dropping it does not weaken the explanatory or the predictive power of the model. In such a case, a scientist would deny that there is any reason to accept the existence of the hypothetical variable in question. As theoretical discourse, talk employing such a variable has no empirical content. For any acceptable scientific model, changes in the value of each of its internal variables must in some situations be detectable in changes in stimulus-response relationships.

A third reason for accepting certain relational attributes as part of the meaning of terms referring to psychological variables is that those terms are theoretical. Unlike observation terms, theoretical terms lack any explicit criteria or test procedures which determine their applicability. Because there are no explicit procedures for determining the applicability of a theoretical term, such a term gains its criteria of application through its links with terms that have such explicit criteria. The theoretical structure gains its meaning from a set of 'correspondence rules' which associate some of the terms with some observable consequences. (See below, p. 129.)

Although this classical account of the meaning of theoretical terms, found in the writings of Carnap, Braithwaite, and Nagel[17] has been faulted on various grounds, it provides us with some insight into the relations between internal variables and stimulus-response relationships. The statement that rein-

[17] See the references cited in Chapter IV, nn. 2 and 3.

forcement consists in drive reduction, for example, lacks empirical content unless we have some way of determining levels of drive independently from determining whether the animal is reinforced. We must be able to infer levels of drive from other kinds of behavioural evidence, such as deprivation or satiation effects. The connection of 'drive' to deprivation and satiation effects gives the concept some empirical content, and with it the claim that reinforcement consists in drive reduction. Now, a sceptic might question the empirical content of the claim that under certain situations, an animal showing deprivation or satiation effects is in a drive state. How can *this* claim have any empirical content unless we have yet further independent grounds for ascribing 'drive'? Obviously, if we could provide such grounds, the sceptic could ask the same question about the empirical content of those claims. Hence, if the theoretical discourse is to have empirical content, its connections to observed stimulus-response relationships in some contexts must be accepted as stipulative. Not all of the connections between an internal variable and observable stimulus-response relationships can be doubted at once; some stipulated relationships are required to give the theoretical term any meaning.

To recapitulate, the terms for an internal variable include in their meaning a relational characterization of the way in which change in that variable will influence other variables and ultimately behaviour. If this were not so then one could not predict how changes in that variable could influence the performance of the system. Furthermore, if the theoretical term is to have meaning in any context, behaviour must in some contexts provide sufficient evidence for ascription of the given internal state. The internal variables are related to behaviour both by conditions of the form:

> In such-and-such situations (holding other relevant internal variables constant), if the animal is in internal state x_i, then it will do such-and-such.

and by conditions of the form:

> In such-and-such situations (holding other relevant internal variables constant), if the animal does such-and-such, then it is in internal state x_i.

In different situations, therefore, behaviour provides a necessary or sufficient condition for inferences to the value of the internal variable. The first is required to make differential predictions from models, the second to give the theoretical terms a use in explanations. The second is also required if one wishes to infer from behavioural evidence that all but one of the internal variables relevant to a behaviour have been controlled; this claim relies both on one's hypothesis of what those other variables are, and on behavioural evidence sufficient to ensure that variations in those factors have been controlled. In this way the claim that one has sufficient controls for a given behavioural phenomenon itself relies on a hypothesis.

The use of the term 'hunger' in explanations of rat behaviour provides an example. Any adequate scientific explanation requires that the evidence for the premisses extend beyond the evidence for the explicandum. In this light, what evidence can show that a rat is hungry, other than the behaviour in question? Length of food deprivation is one kind of evidence. Other factors play a role in the genesis of hunger: the general activity level of the animal during the period of food deprivation, its metabolic activity, the temperature of the environment, and so on. The basic physiological condition in question is energy balance; that is, over a given modulus, does the rat ingest as much energy (in food) as it expends in metabolism and motor activities?

'Hunger' is also related to some consequent variables. The hungry rat will show increased arousal and a heightened general activity level; also, it will learn to perform instrumental reponses in order to obtain food. The harder the rat will work for food, or the greater the aversiveness of stimuli it will undergo in order to obtain it, the hungrier it is presumed to be. The rat will seek food and engage in goal-directed behaviour; it will solve some insight problems in order to obtain food; and stimuli associated with food will themselves be reinforcing.

'Hunger' is a theoretical term defined by its functional role in the genesis of feeding behaviours. The state is defined by certain relational attributes: its relations to physiological conditions (of energy imbalance), to other internal states, and to behaviour. Examples of such relational attributes are the relations between hunger, arousal, reward, need, learning, problem solving, and goal-directed behaviour. A given state is a hunger state only if it shows the proper relations to certain antecedent variables (food deprivation, metabolic activity, and so on) and to certain consequent variables (arousal, goal-directed behaviour, reinforcement). Because some evidence that a rat is hungry is independent of some of the behaviour 'hunger' is invoked to explain, some of these relations must be part of the stipulated truth conditions for 'hunger'. Since they give us grounds for inferring from behaviour in certain situations to the claim that the rat is hungry, it follows that they also give sufficient conditions for a state to be a hunger state.

The stipulative relations need not give an explicit definition of 'hunger'; they need merely link, in certain situations, changes in hunger with changes in behaviour. That is sufficient to tie the theoretical term to observables. Such relational attributes define the conditions which pick out a given state as a hunger state.

Although talk of the 'meaning' of a term may be somewhat woolly, we can fairly say that the truth conditions for any term which denotes an internal psychological variable specify some relations between that variable and its antecedent and consequent variables, so that those relations hold by stipulation for any state which is a value of that variable. Since this is true of every variable in a model, the stipulated relations between a given variable and its antecedents and consequents work outwards to provide certain stipulated relations between the given variable and stimulus-response relations of the system as a whole. A given state is a value of an internal variable in such a model if and only if it bears the appropriate relations to antecedent and consequent variables of that internal variable. If the state in question does not bear the appropriate relations to other parts and states of the system, then it cannot be a value of that internal variable. And since one can infer from behavioural changes to changes in the value of the internal variable, if a given neural state does show the appropriate relations to other internal variables and to stimulus-response effects, then it is a value of the given internal variable.

Perhaps we can clarify this point by considering psychological *functions* instead of psychological variables.

In black box analysis the psychologist asks 'What sort of machinery is required to produce the input-output relationships we observe? What jobs must parts of the system fulfil if the system as a whole is to manifest these relationships?'. Various parts are specified and assigned jobs, 'job' meaning the consequences which the activity of the part has on the activity of other parts and on the system. To specify these jobs is to specify the $functions_c$ of the hypothesized parts.

This implies that if we are to identify a given neural structure as a hypothesized part of the model, we must show that the neural structure fulfils the $function_c$ within the nervous system ascribed to the part in the model. The activity of the structure must bear the appropriate relationships to the activities of other parts of the nervous system; otherwise it cannot fulfil the appropriate $functions_c$.

This point will be expanded below. Now we need only appreciate that a necessary condition for the ascription of a given $function_c$ or $function_g$ to a part of the nervous system is that certain relations hold between the activity of that part and the functioning of other parts, sufficient for us to say that this part does such-and-such a job within the total system, or that this part makes such-and-such a contribution towards maintenance of a *G* state. Both senses of 'function' are defined by the consequences of the activity of a given part on the work of other parts. Hence part of the meaning of terms ascribing psychological $functions_c$ or $functions_g$ is that the structure in question have certain effects, or bear in certain ways upon workings of other parts.

A simple non-psychological example may clarify this point. For a part of a machine to have any $functions_g$, it must be a part in some negative feedback system which maintains some *G* state, and its activity must contribute in some way to maintaining that *G* state. If the channels between the parts of the system are destroyed, then the system itself is destroyed, and its parts can no longer have any $functions_g$. Destruction of the integrity of the system logically entails that its parts no longer have any $functions_g$. Typically, the furnace in a centrally-heated house is controlled by error signals from a thermostat, which fulfils the function$_g$ of error detection in that it contributes in that particular way to the maintenance of a constant temperature in the house. If, however, we cut the wires from the thermostat to the furnace, then the thermostat can no longer fulfil the role of error detector; it cannot play the appropriate functional$_g$ role, since its activity no longer has the required consequences on other parts of the system.

In conclusion, a necessary and sufficient condition for the identification of a neural structure as one of the parts in the model is that its activity have the appropriate relations to activities of other structures. This is a simple consequence of the meaning of the term 'function'.

4. *Localization of Function*

The analysis has not yet touched upon the sorts of experiments in which one 'opens the box'; that is, in which one performs physiological interventions whose intent is to establish the localization of psychological functions. These experiments attempt to isolate the variables involved in particular behavioural phenomena, and then to discover the structures whose input and output corresponds to those variables.

For example, experiments in the physiology of hunger are aimed at discovering the physiological variables relevant to the initiation and termination of feeding. If one accepts that feeding is controlled by a negative feedback device, then one can try to isolate those signals of deviation from homoeostasis which initiate feeding; these are error signals inputting to various behavioural actuator mechanisms. Various physiological conditions, such as stomach contractions (and peripheral stimuli from the gastro-intestinal tract), rate of glucose utilization, fat levels, body temperature signals, and endocrine controls, are suggested to be signals of deviation from homoeostasis.[18] For example, the glucostatic theory proposes that cells in the hypothalamus are differentially sensitive to arterial-venous differences in blood-glucose levels, so that when this difference falls (reflecting low rate of use of blood glucose), a signal is sent to initiate feeding. The fact that insulin injections (leading to a drop in blood-sugar levels) make satiated

[18] Milner, *Physiological Psychology*, pp. 323–7.

animals begin eating, while glucagon injections, leading to a rise in blood-sugar levels, terminate feeding, is evidence relevant to the proposal. It seems likely that glucose levels are monitored and play some role in feeding; if so, they are probably closely tied to endocrine controls (insulin level), a basic set point for body weight, and possibly to some lipostatic control on amount of fat deposited.

Similarly, different signals have been implicated in the termination of feeding. Along with possible glucostatic and lipostatic controls, stomach distension seems to play some role (at least at the limits of the system's capacity), some oral metering is likely to occur, and a change in the reward value of the taste and smell of food seems to occur as food is ingested.[19] Experimental procedures for evaluating the role of these various signals are relatively straightforward. One attempts specifically to eliminate one of the proposed signals, and then simply observes how feeding behaviour changes.

Once one isolates the variables involved, the next step is to discover their sites of action: the mechanisms in the nervous system which are responsible for the behavioural elaboration of the signals. One must suggest a set of links between the proposed signals and the control of feeding behaviour; in terms of the negative feedback model, one needs to isolate the structures in the nervous system corresponding to the behavioural actuator mechanisms. The signals of deviation or satiety are the inputs to these structures, and their output controls the initiation and termination of feeding.

Often what psychologists mean by 'localization of function' is merely that the structure in question is causally relevant to a particular behavioural phenomenon. Often this weak claim is all that one can justifiably make. Suppose, however, that one has a well-developed psychological model. Such a model specifies the functional interrelations of the variables relevant to the organization of the behaviour. Furthermore, it postulates hypothetical structures fulfilling functions$_c$ within the model (see page 64); each part has specified variables as inputs, and its output has specified effects on consequent variables. Localization of function in this context means establishing the neural identity of one of these hypothesized parts in the

[19] Rolls, *The Brain and Reward*, pp. 23–7.

model. The necessary and sufficient conditions for doing this are:

(1) That the input and output pathways of the neural structure correspond to the input and output variables of the box in the model. (The structure must have the appropriate connections to other structures and the neural states of pathways entering and leaving it must be values of the appropriate psychological variables.)

(2) That the job or role which the structure fulfils is the one assigned to the box in the model. (The structure must transform its inputs by the appropriate function$_m$ into its output, and its output must play the appropriate causal role within the workings of the system as a whole.)

To find a structure in the nervous system fulfilling a hypothesized function$_c$ given to a part in a model, one must find a structure which plays the appropriate causal role in the system.

How does one do this? The strategy is to make some physiological intervention, all of whose physiological effects are known, which causes a specific change in the input-output relations of the system. If that behavioural change can be ascribed only to a particular psychological function$_c$ F, then one can conclude that the structures affected by the intervention mediate F.

Typical physiological interventions include lesions, stimulation, drug injections, single-cell recordings, and others such as cooling restricted areas of brain tissue or applying discs of potassium chloride. Alterations in the input-output relations of the system after the intervention ideally reveal the function of the affected structure. If the changes in input-output relations are those predicted by the model and are the specific kinds associated with a given box in the model, then one tentatively suggests that the structure in question fulfils that particular function.

The suggestion must be tentative because such alterations in input-output relations usually can be accounted for in many different ways, and one cannot be sure of the exact changes one has wrought in the nervous system. One must show (1) that only a change in the function in question can account for the input-output changes, and (2) that only because that particular structure has been changed has that function changed. To do this, one must rule out two alternative ways of accounting for the change in stimulus-response relationships:

(1) Some other function performed by parts of the system has been altered, and that accounts for the changes, or
(2) The intervention has affected parts other than the one in question, and it is those effects which have caused the change in function.

Both of these claims are difficult to rule out, because almost any change in stimulus-response relationships can be explained psychologically in different ways, and any intervention has unavoidable effects on parts other than the target structure.

For example, lesions in the region of the lateral hypothalamus of rats produce aphagia: the rat stops eating. For several years it was thought that the lateral hypothalamus was therefore the 'feeding centre' which initiated eating. However, nerves in the nigrostriatal bundle pass very close to the lateral hypothalamus in rats, and in experiments destroying the substantia nigra with drugs, an almost identical effect is produced.[20] Perhaps the lesions destroyed the nigrostriatal bundle, thereby causing the change in behaviour. Differential effects were also found between mid-lateral and far-lateral hypothalamic lesions, suggesting the involvement of nearby fibres running to the globus pallidus.[21]

Experiments on lowered rage thresholds after lesions provide another example of this problem. Often one can account for the experimental findings by supposing that the post-operative scarring involved in any lesion experiment causes a non-physiological stimulus in those centres; and if, as is sometimes likely, those scars cause pain, then the behaviours in question would be produced through a function quite distinct from that supposed by the aggression theorists.

Another example is the role of the ventromedial hypothalamus (VMH) in control of feeding. Lesions in this structure

[20] John F. Marshall, J. Steven Richardson, and Philip Teitelbaum, 'Nigrostriatal bundle damage and the lateral hypothalamic syndrome', *Journal of Comparative and Physiological Psychology*, 87, 5 (Nov. 1974), 808–30.

[21] S. P. Grossman, *Textbook of Physiological Psychology* (New York: John Wiley & Sons, 1967), pp. 360–1.

lead to hyperphagia: the animal eats far more than normal until it attains a body weight of three to four times its normal. One explanation of this finding is that the VMH is the satiety centre which sends the signals for stopping feeding. But there are alternative explanations of why VMH rats get so fat. It has been shown that the metabolic rate of VMH rats is abnormally high and that they synthesize fat during the day (which is abnormal); that their insulin levels are on average six times that of normals, even if the VMH rats are maintained on a normal fixed diet; and that the high insulin level is an effect of VMH lesions.[22] In this case, the specificity of the physiological intervention has not been challenged (there is general agreement that the rats gain weight because they do not have a VMH); instead, an alternative functional change has been proposed which accounts for the new input-output relations.

Other functional changes have been proposed as explanations of why VMH rats become fat. These rats show motivational changes: once they become obese they will not work as hard as ordinary rats for food, even when food-deprived; they will eat more of good-tasting foods than normal rats, yet completely avoid bad-tasting foods even when food-deprived far past the point at which a normal rat would accept such food. Their eating seems to be more controlled by food-stimuli and less controlled by internal stimuli than in ordinary rats; and Grossman proposed that this 'finickiness' explained their obesity. Evidence that such finicky behaviour is shown in VMH rats only after they become obese casts doubt on this interpretation.[23]

In any case, even though VMH rats become obese, it is not necessarily the case that the VMH is responsible for elaborating satiety signals into inhibition of eating behaviour. Perhaps the obesity is a by-product of an endocrine change, perhaps it is

[22] Jacques Le Magnen *et al.*, 'Role of a lipostatic mechanism in regulation by feeding of energy balance in rats', *Journal of Comparative and Physiological Psychology*, 84, 1 (July 1973), 1–23; Gloria A. Tannenbaum, George Paxinos, and Dalbir Bindra, 'Metabolic and endocrine aspects of the ventromedial hypothalamic syndrome in the rat', *Journal of Comparative and Physiological Psychology*, 86, 34 (Mar. 1974), 404–13.

[23] Grossman, *Textbook of Physiological Psychology*, pp. 355–8; Anthony Sclafani and Lawrence Kluge, 'Food motivation and body weight levels in hypothalamic hyperphagic rats', *Journal of Comparative and Physiological Psychology*, 86, 1 (Jan. 1974), 28–46.

due in part to finickiness, or perhaps the body weight set point for the animal has simply been changed by the lesion.

A good way of summarizing the points made so far is by elucidating the relations between the various senses of the word 'function' when a statement of localization of function is taken as successful. We have:

function$_a$: the activity, processes, or operations of a given object: i.e. its functioning.

function$_c$: the consequence of the activity of a given part upon the work of the system as a whole: the job the part does.

function$_m$: a mapping between two sets which assigns only one member in the range to any member in the domain.

function$_g$: the role the activity of a part plays in maintaining a systemic *G* state.

When psychologists talk about the functions required for a black box to show the input-output relationships it does, they often mean the functions$_c$ of its hypothesized parts: the sorts of jobs they do which combine in such a way as to produce the observed regularities. The jobs and relationships of the parts of the hypothesized mechanism account for the function$_m$ relating stimulus variables to response variables. If the model is successfully realized neurophysiologically, then there will exist some parts of the nervous system which fulfil the jobs hypothesized in the model, so that, if the requisite connections are maintained, the functions$_a$ of a given part in the nervous system will fulfil the model-ascribed functions$_c$. (The connections must be maintained, since functions$_c$ are relationally defined, and obtain only as long as do ordinary relationships between the parts of the system.) In that case, those neural structures will correspond to the boxes in the model. But earlier we saw that the boxes transpose their inputs into outputs, so that the output is a function$_m$ of the inputs. Therefore if the mapping is successful, the activity of the neural structure will impose a function$_m$ between the inputs of the structure and its output. Each box in the black box model is assigned a job; in imposing a function$_m$ between inputs and output the box fulfils its function$_c$. The neural structure fulfils its function$_c$ by its ordinary activity (function$_a$) of imposing a function$_m$ between its inputs and its output, and sending that output to the appropriate succeeding structures.

The number of related meanings makes the word 'function' a tricky one, and these senses must be carefully distinguished if talk about functional localization is to be understood. But the very ambiguities of the term perhaps contribute to its usefulness in psychological discourse.

5. *Conditions for Relating a Model to Physiology*

The analysis in the last section suggests the necessary and sufficient conditons for the reduction of a black box model to physiology; each psychological function$_c$ represented by a box in the model must be localized to some structure in the nervous system. The model specifies a collection of parts defined by their functional roles; if one could localize every such part to some structure in the nervous system, satisfying the criteria set in the last section, then one would show that the nervous system is an instance of the sort of mechanism described by the model, and that the nervous system works the way the model claims it works. Obviously one can identify every hypothesized part in the model with some neural structure only if the neural structures are appropriately related to each other and to the variables linking them. Every relation between variables and boxes in the model must also hold for the neural states which are values of the variables and the structures which perform the functions$_c$ specified by the model. What is required is a successful *mapping* of the model on to the nervous system.

To establish a particular claim for localization of function, one must show that the activity of the structure in question bears the appropriate relations to the activity of other structures and to the system as a whole. To ascribe a function to a part is to characterize its relations to other parts; thus claims of localization of function are established only if one shows that the appropriate relations obtain between the part in question and the rest of the system. For example, to show that a neuron is a reward neuron, one must show that it is related to other parts of the system in such a way that its firing increases the probability that an immediately-preceding response will be repeated. Establishing such a localization for every function in the model entails that the model is *functionally isomorphic* to the nervous system. Roughly, this means showing that all of the relations among variables and boxes in the model also hold among the neural states which are values of the variables and the neural structures which fulfil the hypothesized functional roles of the boxes. More precisely, to establish the localization of every function in the model, the following conditions must be met:

(i) For every variable in the model, there is a collection of neural states which are values of that variable.
(ii) For every box in the model there corresponds some structure in the nervous system.
(iii) For every input or output line for a box in the model, there corresponds an input or output pathway to the corresponding structure in the nervous system, such that the states of that pathway are the values of the appropriate variable in the model.
(iv) Each structure in the nervous system corresponding to a given box in the model carries out the transformation on its inputs specified by the model.

The difficult condition is (iii), which can be restated as follows: if we have a class of neural states identified as values of a given variable, they must be related to the appropriate structures in the nervous system—those corresponding to the boxes to which the variable is connected in the model. Equivalently, each structure corresponding to a box must affect and be affected by the appropriate structures in the nervous system. (See Figure 2.8.)

It should be emphasized that neither 'structure' nor 'pathway' necessarily refers to particular structures or pathways which are anatomically distinct and have anatomical names such as 'hippocampus' or 'medial forebrain bundle'. The structures corresponding to boxes of future models may not be congruent with those currently identified by the anatomist's taxonomy. (See below, pp. 101 ff.) The physiological psychologist may need his own taxonomy to pick out the structures and pathways important for his purposes.

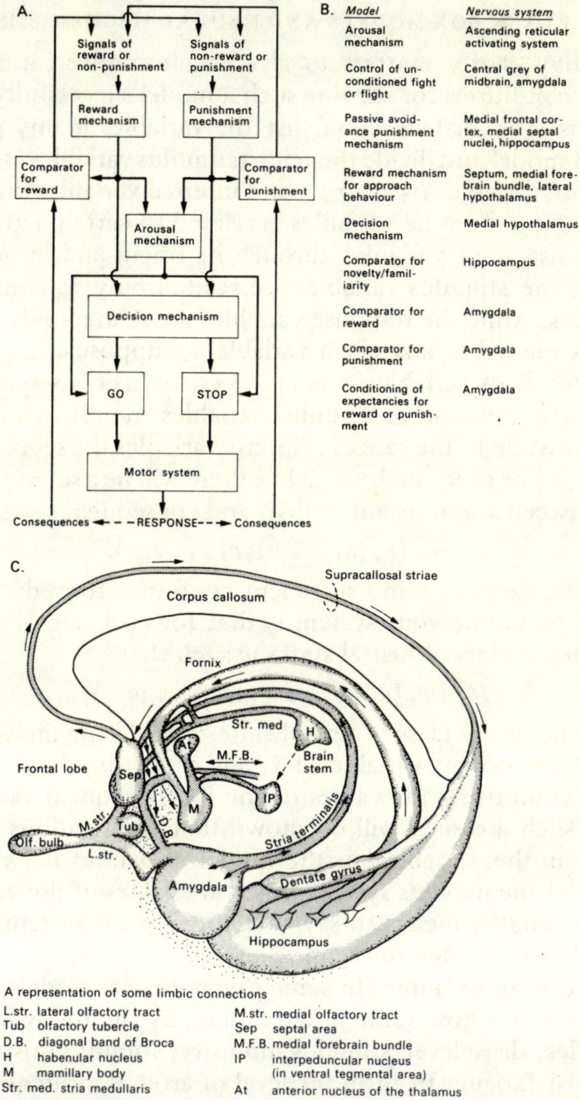

Fig. 2.8. Mapping Gray's model on to the limbic system. ((A) From Gray, *Two-Process Theory*, p. 347. Copyright by Academic Press Inc. (London) Ltd. and reprinted with their permission. (B) From J. A. Gray, *The Psychology of Fear and Stress* (London: Weidenfeld and Nicolson, 1971), p. 217. Reprinted by permission of George Weidenfeld and Nicolson Ltd. (C) From P. D. MacLean, 'Psychosomatic disease and the "visceral brain"', *Psychosomatic Medicine*, 11 (1949), 342. Reprinted by permission of Elsevier North Holland, Inc.)

It may clarify matters to give an alternative formulation of the conditions for relating such a model successfully to the nervous system. One could list the variables in any psychological model and divide them into stimulus variables ($s_1 \ldots s_n$), response variables ($r_1 \ldots r_m$), and internal variables ($x_1 \ldots x_t$). Each of the internal variables is related to certain antecedent and consequent variables through its origin and termination boxes; the stimulus variables are related only to consequent variables, while the response variables are related only to antecedent ones. For any given variable x, suppose $a_1 \ldots a_v$ are its antecedent variables, and $c_1 \ldots c_w$ are its consequent variables. (In the case of stimulus variables the set $\ldots a \ldots$ is empty, while in the case of response variables the set $\ldots c \ldots$ is empty.) For every such variable x there will be a set of relations R_x between x and its antecedent and consequent variables:

$$R_x\ (x, (a_1 \ldots a_v; c_1 \ldots c_w)).$$

Now the necessary and sufficient condition for reducing this model to the nervous system is that for each such variable x one finds a class of neural states w such that

$$R_x\ (w, (w_{a_1} \ldots w_{a_v}; w_{c_1} \ldots w_{c_w}))$$

where w_{a_i} is the class of neural states comprising the values of the ith antecedent variable of x, and w_{c_i} is the class of neural states comprising the values of the ith consequent variable of x. In such a case it will be shown that the relations holding between the variables in the model also hold between the states of the nervous system which are values of the variables. This is what it means to say that the nervous system is functionally isomorphic to the model.

Here is an example. In some psychological models, arousal is a variable whose value is determined by the following other variables: drive level, stimulus intensity, stimulus novelty, and levels of fatigue. In turn, the level of arousal influences other variables such as intensity of response, general alertness, efficiency of learning, and efficiency of performance. If the magnitude of these variables can be measured then the relationships between arousal and its antecedent and consequent variables may be mathematically formalizable, but if not, there will be in any case certain relations R between arousal and its antecedents and consequents:

R(arousal, (drive, stimulus intensity, stimulus novelty, fatigue; response intensity, alertness, efficiency of learning, efficiency of response)).

To show that a certain class of neural states w corresponds to the variable 'arousal', one must pick out other classes of neural states $w_1 \ldots w_8$ corresponding to the list: drive, intensity of stimulus, novelty of stimulus . . . and so on, and show

$$R(w, (w_1 \ldots w_4; w_5 \ldots w_8)).$$

That is, one must find a class of neural states w corresponding to arousal, w_1 corresponding to drive strength, w_2 corresponding to stimulus intensity, w_3 corresponding to stimulus novelty, and so on, such that

$$R(w, (w_1 \ldots w_4; w_5 \ldots w_8)).$$

In this case it would be shown that the nervous system is isomorphic to the psychological model in the sense that the same relations hold between arousal and its antecedent and consequent variables as hold between the class of neural states comprising values of the variable 'arousal' and all those classes of neural states comprising values of the antecedent and consequent variables for 'arousal'.

If, for any variable x_i in the model, one can do this for the relations R_i holding between x_i and its antecedent and consequent variables, then the nervous system has been shown to be an instance of the mechanism abstractly represented in the psychological model.

One further point about the notion of 'functional isomorphism' needs to be made. Two systems are sometimes described as functionally isomorphic whenever the equations describing the relations between their state variables, their inputs, and their outputs are isomorphic: that is, have an identical formal structure but simply use different letters for a given variable. This sense of functional isomorphism is simply an isomorphism between the mathematical functions describing the behaviour of the two systems. The variables in the two sets of equations refer to different sorts of states, yet the formal structure of the two sets is identical.[24]

The functional isomorphism holding between a psychological model and the nervous system must be stronger than this,

[24] See Hempel, 'Aspects of scientific explanation', pp. 435–6.

because such a mapping also requires that the values of the psychological variables be neural states. These identifications are based on congruence of functional role: the given neural state is a value of the internal variable because it bears the appropriate relations to the neural states corresponding to the values of the antecedent and consequent variables of the internal variable in question. The neural state plays the sort of role within the nervous system that the internal variable plays within the model, and so we are justified in identifying that neural state as a value of that variable.

If one could do this for every variable in the model, one would have established that the model is functionally isomorphic to the nervous system, and this would be sufficient to reduce the psychological model to physiology. The connecting principles relating psychological models to the nervous system will be statements of functional ismorphism, of the general form given above. The full implications of such statements will be explored in the remaining chapters.

III
The Logic of Functional Isomorphism

Psychological terms specify the behavioural role or function played by states or processes, while physiological terms specify biophysical features of states or processes of the nervous system. The connecting principles relating these two classes of terms will localize psychological functions to neural structures; the connection between vocabularies is made by claims of functional isomorphism. Because the connecting principles take this form certain possibilities are open in the reduction which would be foreclosed by other sorts of connecting principles. This chapter will examine some of the logical features of connecting principles relating functional terms to structural ones, and the possibilities they open for explanatory reduction.

1. *Some Features of Localization of Function*

The analysis in the previous chapter did not deal with the debate over the question of how 'strictly' or 'loosely' psychological functions are localized in the nervous system. Although I presented an analysis of the conditions necessary and sufficient for claims such as 'The hippocampus is the neural basis of habituation' or 'The orbito-frontal cortex mediates unlearning', I did not clarify the possibilities that, for example, structures other than the hippocampus mediate habituation at other times, or that the orbito-frontal cortex also has jobs other than 'unlearning'. It is an open question how strictly one can pin down a given psychological function to one and only one structure. What are the issues in this argument? The problem is not due simply to the vagueness of physiological terms such as 'hippocampus'—so that one is not sure of the precise boundaries of that structure—nor is it a matter of the terms being somewhat ill-defined, fuzzy-edged, or imprecise. Vagueness and lack of sharp boundaries are features of any class term, but the problem of loose localization of function is due to other features of the terms involved. In order to clarify what it means to say a function is strictly or loosely localized, one must consider some of the logical features of statements of localization of function. It is an open question which of these possibilities will be confirmed by future data; all that is required here is an understanding of the conceptual features of such statements.

A. *Disjunctive*

Because the connecting principles relating psychological to physiological terms relate a functional to a physiological characterization, they may have one feature which has been well discussed by some philosophers.[1] Psychological terms do not specify any intrinsic characteristics or traits of their referents, but instead describe the jobs involved in organizing the input-output relations of the system. A functional characterization of a machine specifies, not the particular physical structure of the machine, but simply the relations between the work of the parts. A psychological variable leaves open the question of the physical features or qualities of the states which are its values.

More formally, a sufficient condition for a state to be a value of a psychological variable is that it bear the appropriate relations to antecedent and consequent variables in the model. If 'x' is the psychological variable, and is related to antecedents $a_1 \ldots a_n$ and consequents $c_1 \ldots c_m$, then our model defines a set of relations $R(x(a_1 \ldots a_n, c_1 \ldots c_m))$. In order to show a neural state w is x, it is sufficient to find other neural states $w_{a_1} \ldots w_{a_n}, w_{c_1} \ldots w_{c_m}$ (fulfilling the appropriate functional roles), and show that $R(w(w_{a_1} \ldots w_{a_n}, w_{c_1} \ldots w_{c_m}))$.

[1] See Hilary Putnam, 'The nature of mental states', in his *Mind, Language and Reality*, Philosophical Papers, Vol. II; Jerry A. Fodor, 'Functional explanation in psychology', in *Readings in the Philosophy of the Social Sciences*, ed. May Brodbeck (New York: Macmillan Co., 1968).

The features or properties of *w* are irrelevant, except in so far as they bear on the relations *w* bears to other states and processes of the system. It is the place *w* occupies in a pattern of relationships which identifies it as an instance of *x*.

Different instances of the same functional state—that is, different values of a given psychological variable—may be instances of different neural states. For example, if a hunger state is just a state which has certain relations to antecedent variables such as food deprivation, and consequent variables such as arousal or reinforcement, then as long as those relations are fulfilled, different instances of a hunger state may be instances of different neural states.

To accommodate this possibility, the connecting principles may need to take a form different from that of general identity statements. One may need to specify the values of a given psychological variable with a disjunctive list of the form

$$\phi_1 \text{ or } \phi_2 \text{ or } \phi_3 \text{ or} \ldots$$

in order to capture the possibility that different neural states fulfil the given psychological function at different times or in different individuals. There may be no single neurophysiological term which picks out just those states picked out by a given psychological term.

A given functional state can occur in systems made up of different materials, or having different structural organizations, as long as the systems show similar functional organizations. The brains of animals of different species may be built somewhat differently, and the structures fulfilling a given function in two species may be non-homologous, yet both species may be accurately described by the same psychological model. Even non-terrestrial organisms made of silicon instead of carbon, or machines made of copper and ceramic, could in some respects share the functional organization of organisms born of mother earth.

Functionalists have often emphasized the possibility that at different times, different neural structures fulfil a given function. It must be stressed that such analyses have established only the logical possibility of that being the case. There is some evidence of functional reorganization within an individual as he or she matures, and maturation includes critical periods after which some functions such as language learning cannot be taken up by different structures. There are cases of congenital absence of structures, with no detected differences in the psychological functioning of the individual concerned.[2] Finally, in cases of disease or injury, remaining tissues sometimes take over the functions of those destroyed. Philosophers use such evidence as support for the claim that in order to specify the values of a particular psychological variable, the mapping principles may take the form of a disjunctive list of neural states 'ϕ_1 or ϕ_2 or . . .'.

In a preliminary way we can say that if a function is *loosely* localized, then different neural states (ϕ_1 or ϕ_2 or . . .) fulfil the particular functional role on different occasions. A *strict* localization of function on the other hand implies that different instances of a given psychological state x are all instances of the same kind of neural state w. Strict localization does not imply that the given function is mediated by a discrete anatomically identifiable structure (such as 'amygdala') or even that the function be localized in some discrete portion of the nervous system; it implies merely that one have some physiological term which picks out just those neural arrangements performing the given job. Those arrangements may be interdigitated with other circuits doing different jobs, they may not correspond to anatomically identified structures, or they may involve certain holistic wave activities in the entire nervous system.[3] As long as one has some term picking out the specific neural mechanisms involved, however, a strict localization can succeed. The option between loose and strict localization of function (which must be settled empirically) is due to the logic of functional concepts.

[2] See M. A. Jeeves, 'Psychological studies of three cases of congenital agnesis of the corpus callosum', in *Functions of the Corpus Callosum*, Ciba Foundation Study Group No. 20, eds. A. V. de Reuck and Ruth Porter (London: J. & A. Churchill Ltd., 1965), p. 76. See also the case cited by Smythies, *Brain Mechanisms and Behaviour*, p. 151.

[3] Karl Pribram in *Languages of the Brain* (Englewood Cliffs, N.J.: Prentice-Hall, 1971) proposes a model of certain modalities of perception and memory involving hologram-like wave interference mechanisms employing the entire cortex. See W. W. Roberts, 'Are hypothalamically motivational mechanisms functionally and anatomically specific?', *Brain, Behaviour and Evolution*, 2 (1969), 317–42 for a review of problems of the interdigitation of circuits and the lack of discrete anatomical boundaries between functional units.

B. *Lesser Scope*

The (potentially) disjunctive form of statements of functional localization is related to their *scope*. The 'scope' of a statement is the collection of individuals over which it is asserted to hold. The scope of a statement of functional localization may range from a particular organism over a particular period of time, up to every individual within a given species, genus, or family. In most stimulation, ablation, or recording studies the intended scope of statements of localization of function is that of normal uninjured adults of a particular sex of a particular species (or laboratory strain). One excludes genetically or developmentally abnormal individuals, injured or diseased individuals, young, immature members of the species, and perhaps members of the species of the opposite sex. Any of these factors may affect the localization of function, and so the scope of the claim must be correspondingly restricted.

The typical strategy when one opens up the black box is to try to delimit a scope within which a given function is always associated with one and only one structure in the nervous system. For example, if one is attempting to discover the structures mediating habituation, one will specify a scope within which that function is always subserved by a given structure (the hippocampus, for example) and only by that structure. Many of the inferences in such studies rely on the assumption that what is true for one individual at one time in the study holds true for others as well, and this requires that the individuals sampled must be ones for which the same statements of functional localization hold. This is why one confines oneself to normal uninjured adults of a particular sex of a particular species, assuming that a localization of function true for one individual in the group will be true for all the others as well. If variations in the data are found, they are ascribed to other factors (side-effects of the physiological intervention or failures in the psychological model) before one suggests that different structures mediate the given function for individuals within that scope.

It is however logically possible that the scope of statements of localization of function be very narrow. Even though a particular structure may always fulfil a particular function in one individual, perhaps different structures do the job in different individuals in the same species. If a single connecting principle applies to every member of a given species, it is possible that functions are localized differently in different species. This opens the possibility that one's psychological model may have wider scope than any particular set of connecting principles relating it to neural mechanisms. For example, a model of classical conditioning may apply to cats, rats, and fishes, although when one maps the model on to neural mechanisms, one may discover that the structures in the fish brain fulfilling a particular function are different from those doing the job in the rat brain. In this case, the model of learning would have greater generality than the connecting principles relating it to neural mechanisms.

Cross-species localizations of function may be possible in some cases although not in others. The generality of the mapping principles may extend to every individual within a phylum, or dwindle to that of a particular individual at a particular time.

The scope of the connecting principles is related to their disjunctive form: laws which hold across species may only loosely specify the structures fulfilling a given function (using a list ϕ_1 or ϕ_2 or . . .), while highly specific localizations of function correspondingly restrict the scope of application. For example, the limbic system in animals of many species subserves motivational functions of reward, drive, and the learning of classical conditioning. However, this is a relatively nonspecific statement of localization of function, for it does not carefully define the precise psychological functions involved, nor does it pick out the particular structures in the limbic system subserving a particular job. If one tries to specify the particular structures involved in unlearning instrumental conditioning, one finds that in rodents this function is subserved by the septal-hippocampal system, while in primates the orbitofrontal cortex takes on that role.[4] Similarly, although it is fair to say that for many species, the hippocampus underlies habituation, if one tries to pick out the particular parts of the hippocampus involved, fine anatomical differences may prevent the more specific connecting principles from applying to both monkeys and men.

[4] Gray, *The Psychology of Fear and Stress,* pp. 206-7.

It is possible that a psychological model has wider scope than the connecting principles relating it to physiology, and that in order to identify the neural mechanisms underlying the psychological functions in the various different species in which the psychological model holds, several different sets of connecting principles will be necessary. In this case the reduced theory would have greater generality than the reducing theory, and one could not derive the laws of the reduced theory from those of the reducing theory. This possibility will be explored further below (pp. 148-9).

C. *Incomplete*

The central problem posed by the possibility of loose localization of function is that one cannot determine a scope within which a particular physiological term (or conjunction of terms) picks out just those neural arrangements fulfilling a given function.[5] I said above that the scope of statements of functional localization is typically taken as normal uninjured adults of a particular sex and species. Yet statements with this scope face problems posed by the evidence concerning individual differences within a species and of functional reorganization within a given individual over time. This evidence may force one to adopt loose localizations of function, so that one would have no physiological term (or conjunction of terms) picking out just those neural arrangements fulfilling a given function in every individual within the scope.

Lashley's experimental technique has been faulted, and few accept the idea of equipotentiality, but the supposition that any one of a number of structures can do a given job may make sense at the neural level. It is possible that different members of a given species are wired differently at a fine anatomical level of detail, and these individual differences may falsify statements of specific localizations of function. For example, in humans the degree to which a cortical structure

[5] One may need a conjunction of terms to localize the function, as in '*CA*3 and *CA*1 cells of the hippocampus'. This specifies that both neural arrangements participate in the given function. If one uses a disjunctive collection of terms, however (as in '*CA*3 cells or *CA*1 cells of the hippocampus'), then the function is not strictly localized, for the connecting principle states two alternative localizations and does not specify which is used at a given time.

on one side is larger than its homologue on the other side varies among individuals; some show asymmetries favouring the right side while others have larger structures on the left. These asymmetries are related to handedness and presumably reflect different localizations of function.[6] The phenomenon of handedness suggests that some functions may be localized in different structures in left-handers than in right-handers. For example, left-handers are much more likely than right-handers to have the right cortical hemisphere dominant for speech functions.[7]

Even more pessimistically, it may be that within a particular individual a function is subserved by different structures at different times. The data concerning functional reorganization during maturation and development,[8] reorganized functions due to genetic abnormalities,[9] and recovery of function after injury or disease, all suggest that even within a particular individual the list of states corresponding to a given psychological variable may be of the form ϕ_1 or ϕ_2 or . . . . The nervous system may be highly redundant, so that any one of a number of different structures performs a given job on different occasions.[10] It may be highly plastic in its functional organization, so that which structure performs a given job is determined in part by activity in other parts of the nervous system.[11] Some structures may fulfil two or more func-

[6] A. M. Galaburda, M. LeMay, T. L. Kemper, and N. Geschwind, 'Right-left asymmetries in the brain', *Science,* 199 (24 Feb. 1978), 852–6.

[7] Milner, *Physiological Psychology,* pp. 253–4.

[8] See A. R. Luria, *The Working Brain* (Harmondsworth, Middx.: Penguin Books, 1973), pp. 74–9 for examples of principles of functional reorganization during ontogeny.

[9] Albinism is a genetic condition resulting in both structural and functional abnormalities in the visual system. See H. Collewijn, B. J. Winterson, and M. F. W. Dubois, 'Optokinetic eye movements in albino rabbits: inversion in anterior visual field', *Science,* 199 (24 Mar. 1978), 1351–3.

[10] Smythies offers this as an explanation for the fact that lesions in both fornices in humans seemingly have few observable effects. See *Brain Mechanisms and Behaviour,* p. 151.

[11] In *Foundations for a Science of Personality* (New York: The Commonwealth Fund, 1941), Andras Angyal argues that the system must use a given part in different functional sub-systems at different times. He postulates setting and shifting mechanisms responsible for organizing collections of parts into functional sub-systems and dissolving and reorganizing them as required, and suggests that many of the symptoms of psychosis are due to failure of integration of the shifting mechanisms. See pp. 135–43, 330–43.

tions,[12] and the system may be economical in other ways so that a simple mapping of structures to functions is precluded.

An example may clarify these points. The reticular formation is a diffuse net of neurons with intermeshed dendrites and long wandering axons proceeding in some cases from the brain stem up into the thalamus.[13] It is generally ascribed the function of behavioural arousal and activation, in particular the cortical desynchronization following a novel or important stimulus. It is in part precisely because the reticular formation is a diffuse and highly redundant net that is reasonable to assign it the function of cortical desynchronization.[14] For this very same reason, however, it is possible that different parts of the reticular formation do the job on different occasions (so that arousal corresponds to a list of different states ϕ_1 or ϕ_2 or . . .), and that one cannot specify all the different combinations of neurons in the reticular formation which do the job at different times (so that the list ϕ_1 or ϕ_2 or . . . could not be completed). Although one can accept the gross localization of function that the reticular formation mediates arousal, once the psychological model becomes more specific, the redundancy, diffuseness, plasticity, or economy of the nervous system may defeat any strict localization of function.

Should this pessimistic prediction be substantiated, the connecting principles relating psychological terms to physiological terms will be *incomplete.* That is, even within a very narrow scope, in order to specify all the neural values of a given psychological variable x, one will need to employ a list of the form

$$\phi_1 \text{ or } \phi_2 \text{ or } \phi_3 \text{ or } \ldots$$

and this list cannot be closed. Not only does one fail to find some scope within which a given function is always mediated by a particular structure, but one may be unable to discover

[12] The hippocampal theta rhythm, for example, may be involved both in habituation and in the process of laying down permanent memories. See Smythies, *Brain Mechanisms and Behaviour,* p. 148.

[13] See Richard Thompson, *Foundations of Physiological Psychology* (New York: Harper & Row, 1967), pp. 435, 459.

[14] The seeming non-specificity and diffuseness of reticular function is probably a reflection of our ignorance, and as physiological knowledge advances it is increasingly differentiated into discrete nuclei each having a precise job. See ibid., p. 435, 441-6.

all of the neural arrangements which on different occasions perform the given job.

There are two different senses in which such connecting principles may be left incomplete. First, localization of function may be so indeterminate, and the nervous system may in some respects be so economical or plastic, that the list of structures which fulfil a given function may be indefinitely long. Even if we assume that only a small number of different neural arrangements mediate a given function within different members of a species, however, there is a second reason why the connecting principles may be incomplete. Loose localization of function may render it technologically impossible for us to *discover* all the different structures which can do the job. The fog of equipotentiality may obscure every trail through the data. For example, suppose that there are three or four different structures which on different occasions mediate the organization of the orienting response in animals of a particular species. How could we discover the particular structures involved? Lesions, stimulation, recording, cooling, and drug injections would at different times have different effects on the orienting response, depending upon which particular neural structure was doing the job at the given time. The data from such experiments would be very confusing and the underlying regularities difficult to detect. Various moves would be made to account for the contradictory evidence. Perhaps individual differences contaminate the results, and so one must restrict oneself to individuals within a narrower scope. Failing that, perhaps the interventions have uncontrolled side-effects at different times, affecting slightly different structures. One can revise one's idea of the psychological functions involved, so that the seemingly different effects on behaviour of a particular intervention in a particular structure are all due to a change in one psychological function. Only if all these moves fail would one entertain the possibility that the structure in question does different things at different times. To sort out the different jobs the structure can fulfil would be very difficult, considering that one would have no idea of the principles determining which structure fulfils the function at a particular time.

If the connecting principles were incomplete in either of

these ways, then one could not use physiological terms to pick out all the values of a given psychological variable. One would not know the physiological features or relations which define the membership requirements of the list ϕ_1 or ϕ_2 or . . . of neural states which are values of the variable. In such a case, the only way one could pick out all the neural states which are values of the given variable would be by specifying their functional role in behaviour. A neural state is a member of the list ϕ_1 or ϕ_2 or . . . if and only if it bears the appropriate relations to the neural states corresponding to the antecedent and consequent variables for x.

It should be emphasized that many of the speculations concerning loose localization of function are merely speculations, and that the data already uncovered make it more probable that localization of function is rather precise. Discovery of novelty neurons in the hippocampus of the cat, bug detectors in the optical nerve of the frog, line detectors in the visual cortex of the cat, and reward neurons in the hypothalamus of the rat all indicate that it may be possible to localize functions to specific arrangements of neurons. Data concerning localization of functions in the limbic system are confusing, but there is no conclusive reason for thinking that limbic structures swap functions over time. Although an appreciation of the logical possibilities involved helps to clarify the logic of reduction, the probability is that for normal uninjured adults of a particular sex and species, a given psychological function is always associated with a particular neural arrangement.

D. *Non-specific*

One implication of the preceding points is that in order to show that the nervous system is functionally isomorphic to a particular psychological model, it is not *necessary* to show that every instance of a psychological state is an instance of the same kind of neural state, nor is it necessary to show that a given function is always carried out by one and only one structure in the nervous system. General identities (linking a psychological variable with a kind of neural state, or a function with one and only one structure) are not required.

The connecting principles may need to take a disjunctive form

$$x = \phi_1 \text{ or } \phi_2 \text{ or } \phi_3 \text{ or} \ldots$$

and these provide at best *particular* identities; so that at a particular time x may be ϕ_1, while at some other time it may be some different state ϕ_2.

This raises the following problem: if functions are only loosely localized, what conditions must be met in order to show that the nervous system is functionally isomorphic to the psychological model? Need one identify, for every instant, the particular neural states playing the role of the psychological variables, and the particular neural structures carrying out a given job? I suggest that the condition on reduction is not so strict. One needs to establish that at every instant the nervous system is functionally isomorphic to the psychological model. This latter condition is not met by discovering the particular identities which obtain at every moment; instead one need have good grounds only for believing that at every instant some particular identity obtains. One need not know exactly which neural state ϕ is the value of x, as long as one has good grounds for believing that there *is* some particular state ϕ which at that time is x. Showing this for every variable and function in the model would be adequate to show that the nervous system is functionally isomorphic to the model.

What could provide good grounds for believing that particular identities obtain? I suggest that physiological psychologists could take certain inductive risks on this question, and that the following three conditions could be considered adequate:

(i) That lesion, stimulation, recording, and drug studies show that the structure in question fulfils the given job (as analysed in Chapter II, Section 4).

(ii) That the structure in question has the requisite neural machinery so that its states could be values of the variable in question.

(iii) That one establish on some particular occasions which neural state ϕ is the value of x, thereby establishing some particular identities.

If one established that the structure in question does the job, that particular neural arrangements in the structure could do the job, and that on some occasions states of those neurons are values of the variable x, then one presumes that at each moment there is some state of the structure which is the value of the given psychological variable.

If conditions of the above form were adequate (and something along these lines would have to do, were functions loosely localized) then the connecting principles mapping the model on to the nervous system will be *non-specific*. By this I mean that those connecting principles would not allow one to establish, for each moment, which particular neural state ϕ is the value of x, or which particular neural arrangement does a given job. According to the above conditions, one can have good grounds for accepting that some particular identities obtain, without knowing what those particular identities are at every instant.

For example, suppose that different parts of the reticular formation mediate arousal and cortical desynchronization at different times. The reticular formation has functions other than arousal (such as organizing the orienting reflex and selective attention, regulating muscle tonus, adjusting the sensitivity of receptors, and participating in sleep regulation)[15] and the particular neurons which do a job may depend upon the other activities carried on by the reticular formation at that time. If this is so then to specify the values of the psychological variable 'arousal' we may need a disjunctive connecting principle of the form '$x = \phi_1$ or ϕ_2 or . . .'. Yet one can still establish that the reticular formation mediates arousal and alerting functions. Stimulation, lesion, and drug experiments show that the structure as a whole has that role in behaviour. Detailed anatomical studies have revealed neurons and neural networks of a form which is well adapted to the function of cortical desynchonization. Finally, recording studies have isolated neurons in the formation whose activity corresponds closely to the processes of arousal, orienting, and habituation. On these grounds one makes the claim that there are particular neurons mediating the alerting function, even though one does not know in every instance which neurons do the job. The connecting principle is *non-specific*.[16]

[15] Ibid., pp. 465–72.

[16] In such a case one would not have physiological terms picking out just those neural circuits fulfilling the given function, and localization of function is therefore loose.

In effect these considerations force a modification in one of the conditions I set as necessary for reducing a black box model to physiology. It is not necessary that one identify precisely the collection of neural states *w* which on different occasions are values of the psychological variable *x* (see p. 80). Instead a well-grounded belief that particular identities obtain at every moment is sufficient. To defeat the reduction, it is not sufficient to show that some instances of a psychological state in the model are not instances of a given kind of neural state. Disproving such general identities is not sufficient; instead one needs to show that *no* neural state at that time fulfils that functional role, that no particular identity statement holds.

E. *Imprecise*

The above four features are ones which statements of functional localization may or may not have; they are mere possibilities opened by the logic of function concepts. Only future data will determine which of them are required in statements of functional localization. In contrast, the final two features characterize every such statement; they are implications of every localization claim. The first is best clarified with an example.

In Gray's model of classical and instrumental conditioning, a septal-hippocampal system is assigned the function of halting behaviour followed by punishment or by the omission of reward. The septal-hippocampal system is linked via the fornix to a medial hypothalamic 'decision' centre; when behaviour is not reinforced, the system 'sends a signal' down the fornix to the decision mechanism, registering the non-occurrence of reward. This in turn causes the decision mechanism to inhibit the behaviour.[17]

Gray's assignments of function are supported by anatomical studies, which show that the structures named have the appropriate interconnections and could do the jobs ascribed; by lesion and stimulation studies, which show that interventions affecting these structures change the performance of the

[17] Gray, *The Psychology of Fear and Stress*, pp. 207, 218.

organism in the appropriate ways; and by drug studies, which show that depression of the septal-hippocampal system retards the learning of conditioned stimuli for punishment or for non-reward. Notice one feature of the connecting principles when this model is mapped on to limbic structures. One could accept that the fornix sends a signal coding the non-occurrence of reward to the medial hypothalamus, even though one does not know the precise form of this signal or the particular neurons in the fornix which are involved. The fornix is a large fibre bundle which probably subserves several different functions, yet it is not necessary to identify the particular states of neurons in the fornix which are involved in the signal transmission in order to localize the function. One need only know that the structures are related to each other in the appropriate ways, that they do their assigned jobs, and that the neurons inside those structures could do those jobs. One need not know exactly *how* the neurons transmit signals of non-reward.

Those signals could be carried by the fornix in many different ways: different neural states could be the relevant ones. Information may be coded in terms of the place in the fornix where spikes occur, or the time at which they occur, or the number or frequency of the spikes.[18] It may be coded in terms of more complicated patterns of time/frequency characteristics, or even in terms of other synaptic events altogether, such as slow-potential changes.[19] One can establish that the fornix conveys signals of non-reward from a comparator for reward to a stop centre, without knowing precisely which sort of neural state is functionally relevant. In such a case one has no physiological terms picking out the signals of non-reward in the fornix; one can only pick them out by specifying their functional role.

Localization of function is *imprecise* in the sense that such claims do not suffice to identify precisely the states of the

[18] The features of neural activity in a particular structure which carry information are largely unknown. See Smythies, *Brain Mechanisms and Behaviour,* pp. 126–7 for a discussion of some of the neural codes possibly involved in cortical EEG activity.

[19] In *Languages of the Brain,* Pribram urges that slow potential changes at synapses constitute a distinct mode of information transfer, a continuous analogue code as opposed to discrete yes–no spikes.

neurons which are the functional$_c$ states of the system.[20] Localization does not necessarily determine the precise modalities of neural function which are relevant psychologically. One need not know exactly *how* the variables are coded in neural processes. This imprecision in the connecting principles is due to the fact that a psychological model specifies the parts inside the black box only in terms of their relations to other parts and to the stimulus-response relationships of the system. To show that states of the fornix are signals of non-reward, one need only show that those states are related to the appropriate antecedent and consequent variables. Many different states of the fornix could satisfy those relational attributes, and the model does not determine which one does the job.

A psychological model will contain *basic parts*—parts which are specified only in terms of their relations to other parts and to the input-output relations of the system as a whole. A basic part will be defined as receiving certain inputs, mapping them on to an output, and sending that output to certain other parts. The internal structure of those parts, or the precise manner in which they carry out their functions, is left unspecified. To be an instance of such a part, a neural structure need only play the appropriate role in the system as a whole; it need not do this in any particular way. Hence variously different neural states could satisfy the given functional description, and a localization of function need not precisely specify the modalities of neural functioning involved.

This may clarify one feature of psychological simulations. Consider the task of building a machine satisfying the functional characterization provided by a psychological model. Certain parts of the machine must correspond to the variables in the model, and the entire collection of parts must produce the requisite input-output relations. The model does not specify how the parts themselves work, and therefore only some of the features of the machine are relevant to the claim of functional isomorphism. It is for this reason that programs

[20] In a negative feedback device only *some* states of the machine are functional states. The others are irrelevant to the feedback system and are non-functional. See Ernest Nagel, 'A formalization of functionalism', in his *Logic without Metaphysics* (Glencoe, Illinois: The Free Press, 1956). It is still an open question which modalities of neural interaction are the functionally relevant ones; that is, which modalities carry information and which do not.

in digital or analogue computers or machines employing diodes and multiselectors can be simulations for behavioural phenomena, even though parts of the simulation work in ways largely different from the ways neurons work. The basic parts of a machine may be something other than neurons, yet the machine will still satisfy the functional characterization provided by the model. The only relevant features are the relations between internal variables, stimulus variables, and response variables; the rest are left unspecified (see Figure 3.1).

It is because a black box model does not specify how its basic parts work that it gives only an imprecise characterization of the states and structures which satisfy it. Until one knows how the basic parts work, one does not know the precise neural states which are functionally relevant.

Perhaps this point about function statements is often missed because it is assumed that black box models can be mapped in only one way on to the nervous system. It is often assumed that:

(1) To every line in the diagram there corresponds some neural pathway (such as the fornix or amygdalo-fugal pathway).
(2) To every box in the diagram there corresponds an anatomically identifiable structure (such as the hippocampus or medial hypothalamus).
(3) The psychological variables correspond to the frequency of spikes of neurons in the pathways.

Although there is a strong temptation to assume that the parts of the model correspond to gross nuclei, and that the functionally relevant states of the nervous system are all those of frequency of axonal discharge, this temptation must be resisted, for it leads to a deep misunderstanding of the role of black box models. The black box model does not carry the implication that it is instantiated in the nervous system in that way.

A very simple Hullian example may clarify this point. In one of Hull's theories,[21] the internal variable ${}_SE_R$ (reaction potential—the probability, given S, that response R will be

[21] This is a slightly simplified version of Hull's 1943 model. See Clark L. Hull, *Principles of Behavior* (New York: D. Appleton-Century, 1943).

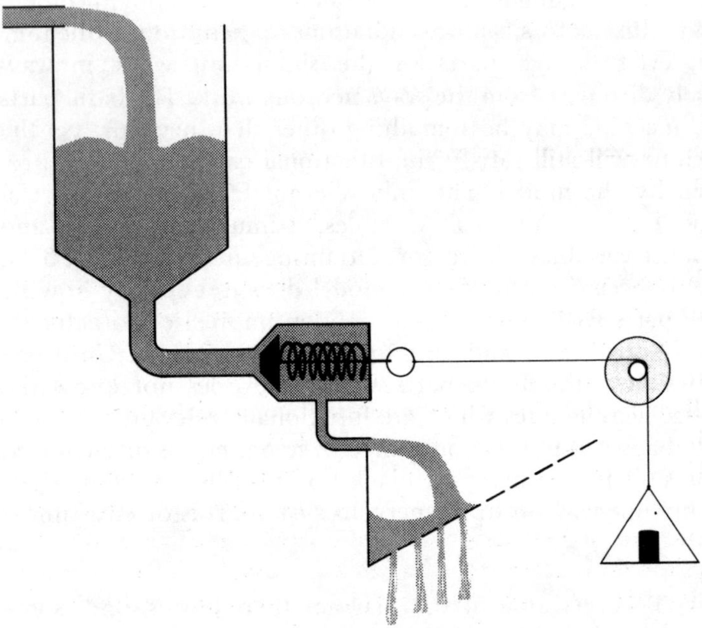

Fig. 3.1. A mechanical model-system: Lorenz's action-specific energy model. Only some of the features of this device are relevant to the claim of functional isomorphism with the nervous system. Lorenz writes: 'the tap supplying a constant flow of liquid represents the endogenous production of action-specific energy; the liquid accumulated in the reservoir represents the amount of this energy which is at the disposal of the organism at a given moment; the elevation attained by its upper level corresponds, at an inverse ratio, to the momentary threshold of the reaction. The cone-valve represents the releasing mechanism, the inhibitory function of the higher centres being symbolized by the spring. The scale-pan which is connected with the valve-shaft by a string acting over a pulley represents the perceptual sector of the releasing mechanism, the weight applied corresponds to the impinging stimulation . . . the internal accumulation of action-specific energy and the external stimulation are both acting in the same direction, both tending to open the valve. . . . The activity itself is represented by the spout discharged from the jet. The intensity of the reaction is symbolized by the distance to which the jet springs, in other words, by the speed of the outflow.' (From Konrad Lorenz, 'The comparative method in studying innate behavioural patterns', *Symposia of the Society for Experimental Biology*, 4 (1950), 221-68. Reprinted by permission of the Society for Experimental Biology and Cambridge University Press.)

made) is claimed to be a function of two other internal variables: habit strength (${}_SH_R$) and drive (D). Diagrammatically:

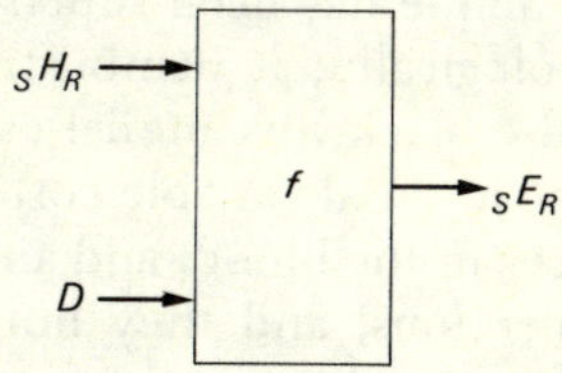

Habit strength was thought to be a function of the number of reinforced performances of R in the presence of S, and drive was thought to be a function of physiological need. Reaction potential is a function of habit strength and drive; specifically, Hull claimed that reaction potential (${}_SE_R$) equalled the product of habit strength (${}_SH_R$) times drive (D).

If we were to map this model on to to the nervous system, need the variables ${}_SH_R$, D, and ${}_SE_R$ all correspond to the frequency of spikes in certain neural pathways? Hull offered a physiological interpretation of this model which proceeded differently. He supposed that habit strength corresponds to the degree of anatomical connections between certain neurons. Needs were thought to change the proportions of certain substances in the blood (glucose levels, saline levels, and so on) and the relative proportions of these substances corresponded to drive. Finally, reaction potential was identified with the degree of conductivity between neurons fired by a stimulus and those leading to a response. Reaction potential would be high if excitation passed easily from one set of neurons to the other, and low otherwise.

The relationships between these variables were explained as follows. When blood containing the drive substance 'bathes the neural structures which constitute the anatomical bases habit (${}_SH_R$), the conductivity of these structures is augmented through lowered resistance . . .'.[22] In this way drive substances would multiply the effect of connections between S neurons and R neurons by lowering their resistance, thereby facilitating the passage of excitation from one to the other. This

[22] See the excerpt from Hull's *Principles of Behavior* in *Motivation*, eds. Dalbir Bindra and Jane Stewart, 2nd edn. (Harmondsworth, Middx.: Penguin Books, 1971), p. 88.

augmented facility to pass excitation corresponded to increased reaction potential.

Although this example has been superseded both psychologically and physiologically, it demonstrates that we must not prejudge how the black box model can be related to the nervous system. One internal variable corresponds to the proportion of substances in the blood, and another to the degree of anatomical connections, and they both act on the same neurons to determine the third internal variable, which is the degree of conductivity of those very same neurons. Frequency of spikes does not play a role. The variables are not states of different pathways, nor do they feed into some anatomically distinct structure.

The black box model presents merely a set of functional relationships between variables; it leaves open the question of which states of the nervous system correspond to those variables. The parts of the mechanism need not be anatomically identifiable structures such as the hippocampus; nor do the variables necessarily correspond to states of anatomical pathways such as the medial forebrain bundle. Strict localization of function does not imply compartmentalized location within some discrete volume of neural tissue, for a circuit doing one job may be interdigitated with other circuits with different functions. All that is required is some physiological term picking out just those neural arrangements doing the job; and those arrangements may not correspond to gross anatomical structures or to discrete volumes in the nervous system.

One feature of the Hullian example that has not been superseded is that the neural change underlying learning (increase in habit strength) is claimed to be a facilitation of connections between neurons. Many theorists explicitly make this identification.[23] Even if we localize the various functions in a model of learning, however, such localization is imprecise, for many different neural changes could conceivably achieve the required facilitation, such as: reverberatory circuits, increased number of synapses, increased size of synapses, increased glial insulation, increased amounts of transmitter substance released at each action potential, increased uptake at the post-synaptic membrane, or increased sensitivity to certain patterns of

[23] Examples are Pavlov, Guthrie, Hebb, early Thorndike, and Milner.

excitation due to RNA changes. One could accept that learning corresponds to facilitated connections between neurons, and successfully map a model of learning mechanisms on to the nervous system, without knowing precisely which one of those biophysical changes is responsible for 'facilitated connections'. The latter problem is that of the neural basis of learning, and is as yet unsolved.

F. *Interconnected*

A final feature of the connecting principles relating functional terms to structural terms is that such principles are *interconnected.* A necessary condition for localizing one function in the model within the nervous system is that the other functions ascribed by the model also be successfully localized.

To understand why this is so, recall the above analysis of isomorphism of function. To show that a given neural state w is a value of a psychological variable x, one must show that the connections of the nervous system are sufficient to allow the state w to bear relations R to other neural states and structures which are also tentatively identified as certain psychological variables or functions. The claim of functional isomorphism is successful if and only if for the given variable x and its antecedent and consequent variables $a_1 \ldots a_n$ and $c_1 \ldots c_m$, if

$$R(x(a_1 \ldots a_n, c_1 \ldots c_m))$$

then for the neural state w,

$$R(w(w_{a_1} \ldots w_{a_n}, w_{c_1} \ldots w_{c_m}))$$

where each w_a or w_c is a class of neural states tentatively identified as the values of those antecedent or consequent variables. But this shows that if one cannot find the appropriate neural states which are values of those antecedent or consequent variables, then the mapping has not been entirely successful, and doubt has been cast on the identification of w states as values of the psychological variable x.

For example, if one finds some class of neural states w which one thinks corresponds to levels of arousal, but for that class w one can discover no connections to some class of neural states w_{c_3}, tentatively identified as the values of variables related to increased efficiency in learning, then the original identification of arousal variables with the class w is rendered dubious. In this way particular identifications rely upon the success of the entire mapping.

Earlier sections have argued that a *sufficient* condition for identifying a neural state w as an instance of psychological state x is that w plays the appropriate sort of functional role within the system, manifesting the appropriate relations to other internal variables and to stimulus and response variables. We must now recognize that for a neural state w to be an instance of psychological state x, it is also a *necessary* condition that w bear the appropriate relations to other states and processes of the system.

For example, suppose that in rats hunger states are those states of the lateral hypothalamus in which the frequency of spikes in neurons Z is q per second. Suppose that we surgically interrupt all of the neural outputs from the lateral hypothalamus, so that activity in the hypothalamus has no effects on the rest of the system. Suppose that our surgery is so neat that it has not completely disturbed the lateral hypothalamus, and that on some occasions the frequency of spikes in neurons Z is q per second. Is the rat hungry on those occasions?

If 'hunger' is a state defined by its relational attributes, then such an operation precludes the possibility that states of the lateral hypothalamus are still hunger states. Hunger is a functional$_g$ state in that it is a variable in a negative feedback system which maintains a G state of long-term energy balance. A necessary condition for the ascription of a functional$_g$ state is that that part be a component of a negative feedback system maintaining a G state; if severance of one of its channels destroys the system, then its parts can no longer manifest functional$_g$ states. If the output pathways from the lateral hypothalamus are severed, then it can no longer be part of a negative feedback system, and its states no longer play a role in the maintenance of any G state. Hence, since hunger states are defined by their functional role, the states of an isolated hypothalamus cannot be hunger states.

It is for this reason that even though one may say that the lateral hypothalamus is the hunger 'centre', one must recognize that talk of 'centres' does not commit one to the claim that states of the isolated hypothalamus are still hunger states. It simply means that states of that structure play a role in the system isomorphic to that of hunger states.

Of course, even on a functionalist analysis, the rat could still become hungry, if other parts of the nervous system took over the functions once subserved by the lateral hypothalamus. However, if we could show that it is physically impossible for the nervous system of the rat to instantiate a class of neural states which bear the appropriate relations to the antecedent and consequent variables for hunger, then we would have shown that the rat can no longer be hungry. This would be similar to showing that the rat cannot see colours, by showing that its visual system does not contain the requisite machinery. It becomes implausible to say that the rat can be hungry in the same way that it is implausible to say that a flatworm or a hydra can expect dinner. The creatures have nervous systems too simple to manifest certain kinds of interconnected states, and therefore they cannot have certain functional states.

In short, showing that a given neural state bears the appropriate relations to other internal variables and to stimulus and response variables is a necessary condition for identifying it as an instance of a psychological state.

In summary, the six features of the connecting principles analysed in this section all derive from the fact that those connecting principles relate a functional characterization of a system to a physiological characterization. Psychological terms do not pick out a class of states by their qualities or features; instead they pick out states by their functional attributes—the role they play in the input-output relations of the system. A state of the system is identified as a psychological state by its relations to other states and structures of the system, and by its contribution to the stimulus-response relationships of the system as a whole.

One implication of this analysis is that a sufficient condition for identifying a neural state as an instance of a given psychological variable is that one show that the neural state plays the appropriate functional role within the system as a whole. This opens the possibility that different instances of a given psychological state may be instances of different neural states, and hence that the connecting principles may need to take a disjunctive form. If in fact many different neural states fulfil a given function, then those connecting principles may take the form of incomplete open lists, with the further consequence that one could not identify, at every particular instant, the specific neural state fulfilling the given role.

The only features which a psychological model provides to identify the neural states which instantiate it are the relations which the given state must bear to other states in the system. That a neural state has those relational attributes is a necessary condition for identifying it as an instance of a given psychological state, and the mapping principles are therefore interconnected. Furthermore, those relational features may be the only criteria available for picking out the members of the list (ϕ_1 or ϕ_2 or . . .) of neural states serving as x on different occasions. In cases of disjunctive or incomplete connecting principles, the only way to pick out all values of the psychological variable is by employing the functional characterization: that class of states which bears such-and-such relation to other internal variables and to stimulus and response variables.

Basic parts in the model are defined only by their relations to other parts and their contribution to the input-output relations of the system as a whole; how the basic parts work is left unspecified. For this reason, even when a function is localized, one does not identify precisely the neural modalities which are psychologically important. Variables can be coded in different ways, and different mechanisms can do the job of the basic parts. This imprecision of localization of function is a consequence of the logic of the concept of 'function'.

2. *Some Peculiarities of this Reduction*

The use of statements of functional localization as connecting principles has consequences for reduction which are not explicit in any of the types of reduction described in Chapter I. Five such peculiarities of the reduction of psychological models to neural mechanisms will be considered in this section.

A. *Substantive Model in Psychology*

According to the model-reduction account, the reducing science provides a substantive model for phenomena described by the reduced science. The first peculiarity of reductionism in psychology is that although neurophysiology is the reducing science, the substantive models are psychological. Analogies between psychological model-systems and the nervous system provide the connecting principles relating the two disciplines (as in 'negative feedback', 'encoder', and 'filter').

This is somewhat odd. Typically the reducing science provides the model-system and the analogies which help us to understand the reduced theory: yet in this case the psychological models help us to understand the way the nervous system works, giving some organization to the otherwise overwhelming and unsystematic mass of neurophysiological detail. Instead of providing an explanatory model-system for data in psychology, neurophysiology seems to provide an accurate instantiation of the psychological model. The nervous system is identified as an instance of the class of machines described by the psychological model. In typical cases of model reduction the opposite happens, for systems described by the *reduced* science are identified as instances of the hypothetical mechanisms described by the *reducing* science.

This feature of reduction in psychology is not particularly important, since, to anticipate slightly, both psychology and physiology aim at producing a unified account of the activity of the nervous system. The peculiarities mentioned here all stem from the fact that proposing a psychological model seems to be the best way to find a unified account. However, once the neurophysiological identifications are made, and we find neural structures which carry out all the jobs assigned to hypothetical parts in our model-system, we can as accurately call that model physiological as psychological, for it will suffice to explain the behavioural regularities, while it rests in part on data originating in the reducing science.

B. *Empirical Support for Mapping Principles*

A rather tricky logical feature of statements of functional isomorphism is that the empirical support for any one such statement rests on the collective success of a large number of such claims. Any particular identification is only as successful as the whole mapping, for only if the entire mapping is more or less successful can one claim with any justification that a given neural state fulfils a functional role isomorphic to the one assigned to a given psychological variable. This is an implication of the fact that statements of localization of function are interconnected (see p. 105) and that a psychological model must be mapped as a whole on to the nervous system.

It is difficult to avoid overstating this point, since in actual experiments, failure in minor details of the mapping certainly does not overturn major identifications. For example, according to the Gray-Smith model, the amygdala is the mechanism which associates punishing or rewarding stimuli to initially neutral ones, and compares received rewards or punishments with expected ones.[24] Now if it could be shown that the amygdala does not receive any inputs from the mechanism which is otherwise successfully identified as the reward mechanism, then the amygdala could not fulfil those functions, and the identification of the neural structures underlying both functions would be cast in doubt (see Figure 3.2). Some adjustment in the mapping or in the model would be required. The model must be successfully mapped in all its aspects to neural mechanisms, or else all of the identifications are rendered dubious. If, to continue the example, one found some structure other than the amygdala which had the requisite connections with the hypothesized reward mechanism—plus those needed to other structures—the new identification of the comparator for reward would also strengthen the localization of the reward mechanism.

It is an exaggeration to say that *all* the functions ascribed to hypothetical parts of a mechanism must be mapped on to the nervous system together, or that any particular failure in the mapping defeats the entire set of mapping principles. The mapping need not be perfect or complete to be generally

[24] Gray, *The Psychology of Fear and Stress*, p. 209.

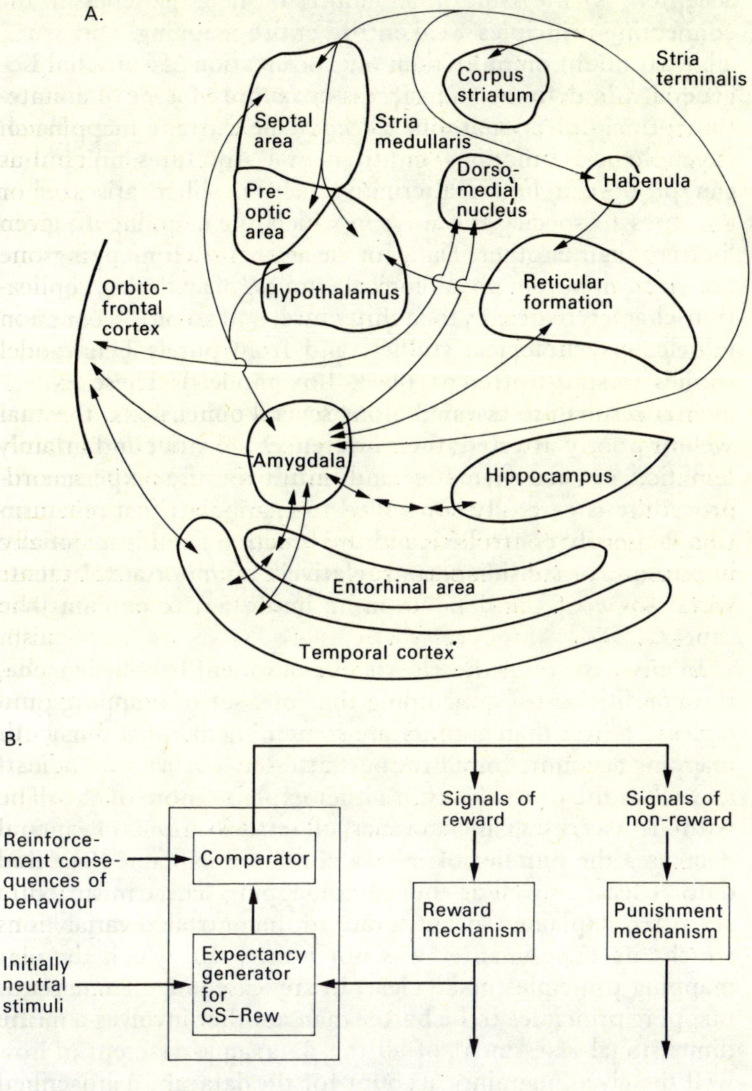

Fig. 3.2. The comparator for reward and the connections of the amygdala. ((A) derived from J. R. Smythies, *Brain Mechanisms and Behaviour,* 2nd edn. (Oxford: Blackwell Scientific Publications, 1970), p. 30. Reprinted by permission of Blackwell Scientific Publications Ltd. (B) derived from text of Gray, *The Psychology of Fear and Stress*, pp. 187–8, 208–9; Reprinted by permission of George Weidenfeld & Nicolson Ltd.)

accepted. At the same time, failures or incompleteness in the connecting principles weaken the entire mapping, and generally cast doubt on more than one localization of function. The general rule defining the success or failure of a set of connecting principles is that one accepts the current mapping of psychological functions on to neural structures unless one can propose a better alternative. The problem arises when one tries to specify the ways in which one mapping is judged 'better' than another. Data for or against such mappings are drawn from purely physiological studies (of anatomy, conduction characteristics, cyto-architecture, and so on), from physiological-psychological studies, and from purely behavioural studies (testing different black box models). These experimental results are assessed along several dimensions: they are well or poorly attested; their inferences are strict and unproblematic, or unsystematic and intuitive; the experimental procedure is carefully controlled to manipulate just one function, or poorly controlled; and the function itself is major and important, or subsidiary and relatively unimportant. In these ways, some of the data are more important to explain than others.

In any case, after these sorts of assessment have been made, the conditions for concluding that one set of mapping principles is better than another are something like this. First, the mapping accounts for all the best-attested localizations at least as well as the other. Next, it either explains more of the data, without increasing the number of serious mismatches; or it decreases the number of mismatches, and explains the other data at least as well as the other mapping. These mismatches are often explained away as due to uncontrolled variables of which the experimenter was not aware, and which the new mapping principles make clear. In any case judging one set of mapping principles to be better than another involves a multidimensional assessment of all the data; an assessment of how well the given mappings account for the data; and a judgement of the importance of failures or successes in those mappings, of the rigour of the explanations they provide, and of the plausibility of their attempts to explain away mismatches or contradictory data. In this way particular mismatches do not defeat the entire mapping (although they do weaken it); and perhaps eventually they make one seek a new set of mapping principles which will do better.

C. *Unified Model*

There is one rather strong conditon on reduction in psychology. If it can be shown that the sort of machinery required by the psychological model does not exist in the nervous system, then the psychological model is treated as disconfirmed. For example, if a model of visual perception requires that the optic nerve have an information-carrying capacity of 10^9 bits per second, and it can be shown that the optic nerve has a channel capacity of only 10^8 bits per second, then that model of visual perception has been disconfirmed. As Gray says: 'Any account of behaviour which does not agree with the knowledge of the nervous and endocrine systems which has been gained from the direct study of physiology *must* be wrong.'[25]

The oddity of this last point is plain: theories in statistical mechanics could not in this way disconfirm those of thermodynamics. The reason, I suggest, is that the sciences of psychology *and* physiology both contribute towards the project of reduction. Both strive to achieve a unified physiological-psychological model of the functioning of the nervous system. Physiologists and psychologists ask different, but related sorts of questions; their activities are complementary, aimed at a unified theory; and both disciplines make contributions to that theory. We can find many examples of physiological research contributing to shifts in psychological theorizing. Consider the simple expedient of inserting an oesophageal fistula in an animal so that it either 'sham eats' or can inject food directly into its stomach by pressing a bar. Dogs will sham eat several times their normal meal size, which shows (a) that oral-pharyngeal stimuli are reinforcing for eating and (b) that the satiety factors are largely alimentary.[26] Similarly, rats have difficulty learning to inject foodstuffs directly into their stomachs, but preloads do affect the amount they will self-inject.[27] Alimentary factors are therefore not as reinforcing

[25] Ibid., p. 181.
[26] Milner, *Physiological Psychology*, p. 326.
[27] Rolls, *The Brain and Reward*, pp. 21–2.

as oral-pharyngeal ones, but are more relevant to satiety. The independence of satiety and reinforcement factors effectively refutes the psychological theory that reinforcement consists in need reduction, for rats will sham eat even though no food reaches their stomach.

One sense in which psychological and physiological understandings of the nervous system will (ideally) unify is that psychologists postulate a pattern of relationships between various internal states in the system, explaining its stimulus-response regularities, while physiologists identify the neurophysiological states which satisfy those patterns of relationships. If one finds that physiological processes do not satisfy the structure of relations postulated by a psychological model, then one abandons the psychological model. This is why the sham-eating experiments are damaging for the view that reinforcement consists in need-reduction.

Our understanding of the nervous system need not stop with the discovery that various neural states and processes stand in various complicated relations to each other, thereby fulfilling the functional roles assigned to internal states in the psychological model. The physiologist wants to *explain* those regularities in the internal workings of the nervous system. Why, for example, is reticular activation co-ordinated with hippocampal theta waves? In asking such questions, the physiologist wants to clarify the $function_c$ of various parts in the nervous system, their ordinary effects on activity and structures in other parts. The physiologist tries to answer such questions by considering structures in the nervous system as black boxes whose input-output regularities are known but whose internal working is unclear. In a way to be further clarified below (p. 120), both the psychologist and the physiologist are interested in the functions and functioning of the nervous system: the psychologist from the perspective of the role its states and structures play in behaviour, and the physiologist from the perspective of explaining regularities in the activities and processes in the system. A second way, therefore, of characterizing a unity of psychological and physiological knowledge is that psychological models ideally will become detailed enough so that the functions a psychologist ascribes to hypothetical parts within the black box are jobs which the physiologist can plausibly assign to collections of neurons. When it is established that particular neural arrangements *do* carry out the jobs assigned to hypothetical parts of the psychological model, then reduction has been achieved. In this way physiological knowledge of *how* structures in the nervous system work (and produce the observed patterns of neural activity) ideally connects with psychological knowledge of the role which those internal mechanisms play in behaviour.

Physiology remains the reducing science in the sense that it explains certain regularities which psychological models take for granted: the patterns of relationships holding among internal states and processes of the system. These patterns of activity are explained by showing how the structures of the nervous system work.

I said above (p. 100) that a psychological model has certain basic parts which are characterized only in terms of their functional role. A simulation of the psychological model is successful if the machine has parts satisfying the functional characterizations of the basic parts of the model. Only some of the states of the machine are important to the claim of functional isomorphism; in particular, since the model does not specify how its basic parts work, variously different mechanical interpretations of the model are possible, in which different sorts of basic parts all meet the same functional characterization.

A third way of characterizing the unity of psychological and physiological accounts is that psychological models will develop to the point where their basic parts are neurons or collections of neuron-like elements (see Figure 3.3). The functional characterization of the basic parts will become more and more specific until the basic parts can only be elements which work in very much the way physiologists understand neurons to work. In that case the model could not be instantiated in a machine whose elements work much differently from neurons. Furthermore, the basic parts of the model would be elements whose functioning can be explained by physiology.

To forestall confusion, it should be pointed out that the connections between the disciplines can be made at different levels of detail. Even a gross localization of general functions

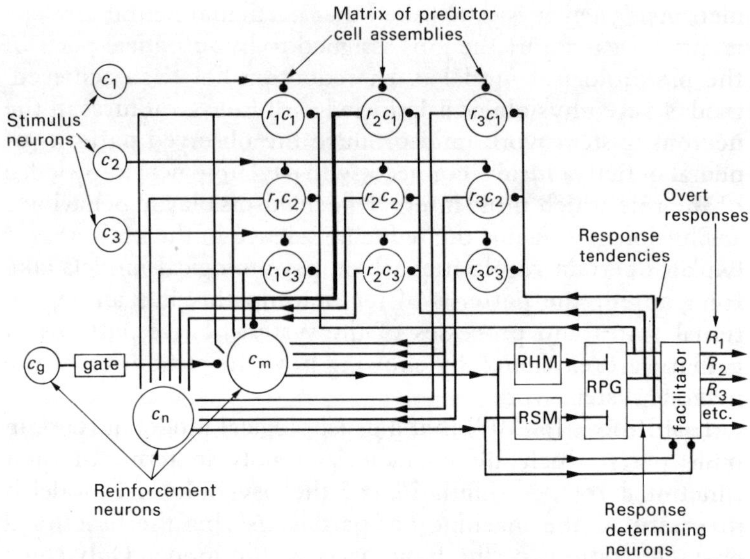

Fig. 3.3. A physiological model of expectancies. The $r_i c_j$ neurons are predictor cell assemblies which 'reverberate' when response determining neurons R_1 fire at the same time as neurons fired by a particular stimulus c_j. The $r_2 c_1$ cell assembly has fired shortly before the c_m neuron, thereby acquiring a connection with it, marked by the heavy line. This represents the expectancy that the response R_2 made in the presence of stimulus c_1 will lead to firing of c_m. Firing of the $r_2 c_1$ cell assembly thenceforth leads to an increased chance that c_m will fire. Reinforcement (from c_g) in this model is thought to activate a response hold mechanism (RHM) which turns a response tendency (generated by the response pattern generator RPG) into an overt response. Once the expectancy $c_1 R_2 c_m$ is learned, therefore, in the presence of c_1, occurrence of the response tendency R_2 will fire c_m, thereby activating the RHM and allowing R_2 to emerge as an overt response. Reproduced by arrangement with Holt, Rinehart and Winston, New York, from *Physiological Psychology*, by Peter M. Milner, © 1970, p. 405.

(such as reward, comparator, or behavioural inhibition system) to macrostructures in the nervous system (such as the hypothalamus, hippocampus, or septal-hippocampal system) connects the psychological model to physiology in so far as it relies upon physiological knowledge of the gross anatomy and interconnections of those structures. When one considers the internal activity of the comparator or the unlearning mechanism, one relies upon knowledge of the basic cyto-architecture and internal organization of structures such as the hippocampus or orbito-frontal cortex. Finally, network models rely upon knowledge of the physiology of particular neurons, and some models of memory make claims about the physiological functioning of the synapses.

We can analogize the relations between the physiological and the psychological endeavours to the building of an arch. We begin two spans on the ground, and hope that they will meet appropriately overhead. The ultimate connection between the two spans is important for supporting the entire structure. Although here and there in the spiderwork the two arches have been connected, at other places we have only a rough indication of where the two might meet, and at yet other places construction is confused and we have no idea where it is heading. Workers on either span constantly adjust their efforts to take account of the construction on the other side. These mutual adjustments are made with the explicit aim of eventually joining the two together.

D. *Basic Parts as Black Boxes*

One consequence of mapping a psychological model on to the nervous system is that one can then proceed to treat any basic part of the model as a black box and ask how it produces the output it does, given its input. A psychological model will specify a basic part by defining a set of relationships between antecedent internal variables $a_1 \ldots a_n$ and x, so that $x = f(a_1 \ldots a_n)$. Diagrammatically:

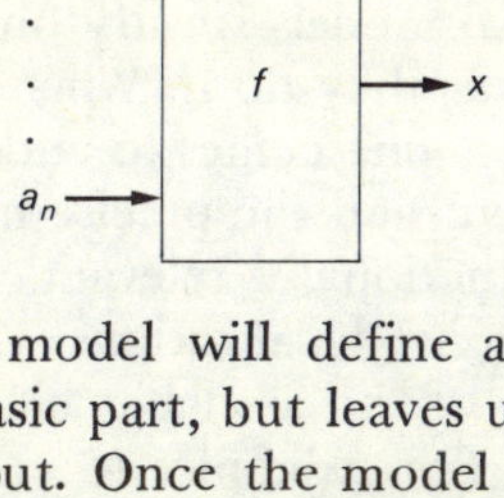

The psychological model will define a set of input-output relations for the basic part, but leaves unspecified how such relations come about. Once the model is mapped on to the nervous system, however, one can proceed to ask how the basic parts do their jobs and produce their input-output relations.

For example, in Sokolov's model of habituation, a mechanism which compares current stimuli with a representation of expected stimuli either triggers arousal mechanisms (if there is a mismatch) or inhibits the orienting reflex (if there is a match). The comparator job has been assigned to the hippocampus. Once this mapping has been made, one can proceed to ask how the hippocampus does the job of comparing incoming stimuli with expected stimuli (see Figure 3.4). One needs to specify the neural inputs corresponding to incoming stimuli, and those corresponding to extrapolated expected stimuli; the mechanism in the hippocampus which compares the two; and the precise output pathways to the mechanisms mediating the orienting response or habituation. When one inspects the neural anatomy and neural functioning of parts of the hippocampus, one finds neural circuits in the structure which could be given such jobs, yet others whose role is unclear. How do the neurons in the hippocampus work together to achieve the comparator function? One must ask questions such as: In what way are incoming stimuli coded? How are expectancies stored? Is the comparison made on the basis of impulse place, frequency, time, or some other feature? What form of output corresponds to a signal of mismatch? Do the same neurons always do the comparator job? And so on.

One considers the psychologically-defined basic part as itself a black box, with internal parts co-operating in an unknown fashion to produce its input-output relations. The parts are neurons and the input-output relations are patterns of physiological states, but the basic logic of black box analysis is otherwise the same as in psychological model building.

The imprecision of functional descriptions derives from the fact that psychological models specify only what their basic parts do, and not *how* they do it. When one discovers how the parts do their jobs, one comes to understand the sorts of neural coding involved, and the precise modalities of neural activity which are functionally relevant. It follows that the imprecision of functional localizations is lessened when one treats the basic parts of the model as black boxes and finds out how they work. For example, to claim that the neural

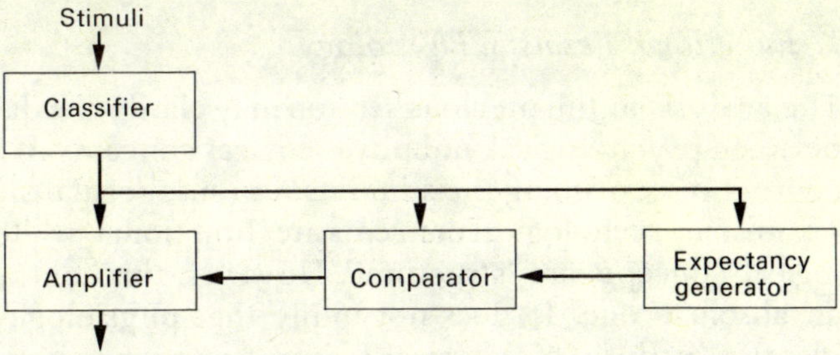

A. The place of the comparator in Sokolov's model

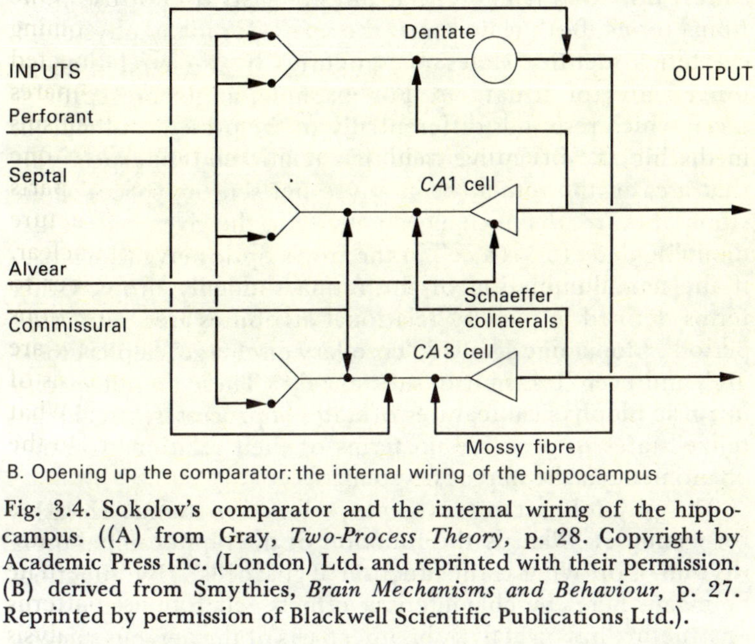

B. Opening up the comparator: the internal wiring of the hippocampus

Fig. 3.4. Sokolov's comparator and the internal wiring of the hippocampus. ((A) from Gray, *Two-Process Theory*, p. 28. Copyright by Academic Press Inc. (London) Ltd. and reprinted with their permission. (B) derived from Smythies, *Brain Mechanisms and Behaviour*, p. 27. Reprinted by permission of Blackwell Scientific Publications Ltd.).

change underlying learning is one of facilitated connections between neurons identifies the neural states involved only imprecisely. As mentioned above, facilitation could result from increased numbers of synapses, larger synapses, glial insulation, RNA changes in a cell, increased transmitter substance, and so on. This imprecision is only rectified when the problem of the neural basis of learning is solved. When one understands *how* facilitated connections come about then one can specify more precisely the neural modalities which are crucial.

E. *Functional Terms in Physiology*

The analysis in the previous section may clarify the distinction between psychological and physiological concepts. Within the context of explaining the stimulus-response regularities of the organism, psychological concepts are 'functional' while physiological concepts are 'structural'. However, this contrast is not an absolute one. It does not imply that physiological terms identify intrinsic biophysical features or properties of neural states; nor does it imply that physiologists do not use functional terms. Both claims miss the mark, for many physiological terms identify states or structures by their relations to other states or structures. For example, a 'glucoreceptor' is a cell which responds differentially to the presence of glucose in the blood. 'Receptive field' is a relational term, specifying that area in the retina (or in any other sensory space) where stimulation results in changed activity in the given cell. A 'net dimming detector' is a cell in the frog's optic nerve which fires if the net illumination of the retina suddenly drops. Other terms defined in part by relational attributes are: 'refractory period', 'dopamine agonist', 'corollary discharge', 'alpha blocking', and even 'transmitter substance'.[28] These do not specify intrinsic biophysical features of states or processes, but characterize states or processes in terms of their relations to other phenomena in the nervous system.

The physiologist often employs function terms. He asks, for example: 'What is the function of the hippocampal theta rhythm?', or 'What is the function of glial cells?'. By 'function' he means here the characteristic effects which the given state or structure has on activity in other parts of the nervous system. The physiological function of the hippocampal theta ryhthm would be clarified by elucidating the role it plays in cortical desynchronization, its inhibition of the reticular formation, and in general its characteristic effects on other structures and processes in the nervous system. To explain how synapses work, how limbic structures influence one another, or why

[28] 'Transmitter substance' picks out a class of compounds by their functional role. A decade ago it was unclear whether there was such a thing as a transmitter substance, for the debate between chemical and electrical hypotheses of nerve action at the synapse was unsettled. See Thompson, *Foundations of Physiological Psychology*, pp. 192–5.

REM activity is accompanied by pontine-geniculo-occipital (PGO) spikes, the physiologist employs functional concepts.

One use of functional terms in physiology is to name certain characteristic relationships between states or processes in the nervous system. Terms such as 'excite', 'stimulate', 'facilitate', or 'inhibit' can label relations between physiological states, so that a physiological variable a_1 'stimulates' x if increased activity of a_1 causes increased activity of x. A second use of functional terms in physiology is to name the jobs or roles of parts of the nervous system. They are used in explaining how the structures in the nervous system produce their input-output regularities. Functional terms in physiology are important when one investigates the basic parts to explain why neural activity shows the regularities it does. To explain why a physiological state a_1 leads to increased activity of x, one needs to understand how the intervening neural arrangement works. That in turn involves postulating internal parts working together in certain ways to produce the input-output relations. Those internal parts are assigned jobs or roles—and there functional terms in physiology find a second use.

Functional terms in physiology are distinct from those in psychology because the relevant input-output relations are of a different order. One sort of term is used to explain input-output relations of parts of the nervous system, while the other is used in the context of stimulus-response relationships. The relevant black box and input-output relations are defined in different contexts.

For example, 'habit strength' is functionally defined in terms of its role in producing stimulus-response relationships. It is connected to certain antecedent variables which determine its value and to certain consequent variables upon which it has characteristic effects. 'Habit strength' could be instantiated in systems made of variously different materials and built in different ways. Although Hull supposed that habit strength corresponded to facilitated connections between certain neurons, certainly other physiological mechanisms could satisfy the relational attributes ascribed by the term.

'Facilitate' is itself a functional term, although it is defined in terms of physiological regularities. If we claim some structure facilitates the connections between *S* neurons and *R* neurons, we are describing a job which that structure performs, a set of characteristic effects it has on other states and processes. 'Habit strength' could be instantiated in systems built differently and made of different materials; similarly, 'facilitated connections' could be achieved by variously different biophysical changes.

The main difference between these two terms is that while 'habit strength' could be instantiated in systems which do not contain neuron-like elements, 'facilitated connections' could not. Functional terms in psychology are defined in terms of relationships between stimuli and behaviour. The black box comprises the entire organism, and one postulates various internal variables and parts in the black box to account for the stimulus-response relationships. In contrast, functional terms in physiology are defined in terms of relations between states within the nervous system. The black box comprises some part of the nervous system, and one describes the characteristic relationships between its inputs and outputs, or the characteristic effects of its activity on other structures.

Psychology and physiology define their function terms for different domains, yet the co-operation between the two disciplines is clear: if all goes well, the function terms of the psychologist postulate a structure of relationships satisfied by physiological states, and the physiological function terms are used to explain the regularities holding between different neural states. The physiologist explains how the basic parts in the psychological model generate their input-output relations.

The distinction between functional and structural terms is not an absolute one, but is dependent on one's context of description; that is, dependent on how one defines the relevant black box, its input-output relations, and its basic parts. Terms specifying the input-output relations of a system or the jobs its internal parts must do are functional terms relative to those used to identify the mechanisms fulfilling those jobs, which are structural terms. If the black box is defined psychologically, physiology thus provides it with a structural description. At the same time, however, one can consider the basic parts of the psychological model as themselves black boxes, and accept that the patterns of relationships between internal states in the system are in need of explanation. For these tasks the physiologist employs functional terms. The functions in question are defined in different contexts, yet they may ultimately connect in a unified model of the functioning of the nervous system.

IV
Implications for some Philosophical Accounts of Reduction

Perhaps the best way to clarify and complete the analysis of explanatory reduction in psychology offered in the previous two chapters is to compare it with the various philosophical accounts of reduction presented in Chapter I. Those accounts provide certain lists of features found in various types of reduction in the history of science, and my goal is to specify which of those features are shared by reduction in psychology, and which are not. Perhaps the mapping of psychological models on to the nervous system does not fall neatly into any one of the three types I presented, but displays features of all of them. In this chapter, I shall consider the Identity theory and model-reduction, and then consider some functionalist arguments against the possibility of reduction. In the final chapter I shall consider eliminative-reduction.

1. *Prospects for General Identities*

It will be recalled that a general identity statement is one which asserts an identity between the *classes* 'psychological state *x*' and 'neural state *w*', so that every instance of a psychological state *x* is an instance of a neural state *w*. We could also call such statements 'type-type identity laws', for they assert that different instances of a given psychological state *x* are all instances of a particular kind of neural state *w*.[1]

According to the Identity theory (that is, the Identity version of derivation-reduction), such general identities are crucial to reduction, for they serve as the connecting principles rela-

[1] See Jerry A Fodor, *The Language of Thought* (Hassocks, Sussex: The Harvester Press, 1976), p. 14.

ting psychological terms to physiological terms. As explained above (see pp. 30-3), the basic feature of this type of reduction is that the reducing theory explains all the laws of the reduced theory, and to do this the terms in the two theories must be related to each other sufficiently for the reduced theory to be *logically derived* from the reducing theory. Although the connecting principles required for such a logical derivation need not be identity claims, Identity theorists claim that the connecting principles will show that a given psychological term '*x*' and a neurophysiological term '*w*' are precisely coextensive, both referring to the same class of states. Physiological terms will pick out just those states which are instances of the given psychological states. In that way our knowledge of physiological regularities could explain the psychological regularities.

Although it is *possible* (and perhaps likely) that psychological terms and neurophysiological terms will be related by general identity statements, the analysis in the previous two chapters indicates that reduction could be achieved *without* such general identities. Reduction is achieved if one establishes that the nervous system is functionally isomorphic to the psychological model, and this can be done even without general identities linking every psychological term '*x*' to a precisely coextensive physiological term '*w*'. For example, because psychological terms pick out their referents in terms of their functional role, it is possible that different instances of a given psychological state *x* are instances of different neural states (ϕ_1 or ϕ_2 or ϕ_3 . . .), so that different kinds of neural states act at different times as values of the given psychological variable. It was argued that reduction can be achieved even though connecting principles of this disjunctive form are required, and this implies that type-type identities are not necessary for reduction.

It may be objected that this is a merely terminological problem, based on an accident of the taxonomies employed by psychologists and physiologists. In particular, it will be argued, if we have a collection of neural states

$$\phi_1 \text{ or } \phi_2 \text{ or } \ldots \text{ or } \phi_n$$

which at different times serve as values of the variable *x*, why not simply take the union of the above sets to define a new physiological term ϕ_m, by definition coextensive with x? Would we not then have a general identity $\phi_m = x$?

This will not do, however. The Identity theorist requires a physiological term which picks out the very same states as the psychological term 'x', and the new term 'ϕ_m' picks out too much. At any one time only one state out of the collection $(\phi_1, \phi_2, \ldots \phi_n)$ serves as the psychological state x, while the others may be occupied in other functions. The union of those classes therefore will not pick out just the values of the variable x; it will pick out other psychological states as well. To get a general identity, we also need to specify *which one* of those neural states is at any given time serving as x. It was argued above, however, that reduction can be achieved with *non-specific* connecting principles, which do not allow us to specify which neural state (out of the collection $\phi_1, \phi_2, \ldots \phi_n$) does the job at any particular time. One may have good grounds for believing that some particular identity obtains, without knowing which one does. In this way, reduction can be achieved even though one has not picked out just those physiological states which at different times are instances of a given psychological state.

There are other possible features of the connecting principles which show that reduction can be achieved without physiological terms picking out just the referents of the psychological terms in the model. The connecting principles may be incomplete, and they will certainly be somewhat imprecise. For these reasons one may not have any physiological criteria of membership of the list 'ϕ_1 or ϕ_2 or . . .' of instances of x. In that case one could find no neural microfeatures which suffice to pick out just those physiological states which are psychological states. Perhaps the only way they can be identified is by referring to their functional role; that is, that they bear such-and-such relations to stimulus variables, other internal variables, and response variables. General identities are not necessary for reduction to succeed.

I said above (p. 91) that a function is *strictly* localized if and only if within some scope one finds a physiological term (or conjunction of terms) which picks out just those neural arrangements doing the given job. A function is *loosely* localized if and only if it is not strictly localized. A general identity statement shows that within some scope a physiological term (or conjunction of terms) picks out just those states or structures picked out by a psychological term. The problem of finding general identities linking psychological and physiological terms is therefore the very same problem as that of establishing a strict localization of psychological functions. If functions are in fact loosely localized (so that the mapping principles need to take disjunctive, non-specific, or incomplete forms) then one will not find some physiological term picking out just those neural arrangements performing the function, and therefore general identities will not be found.

It should be emphasized that none of those arguments establish that general identities *will* in fact *not* be found. The point is simply that it is *possible* that functions are not strictly localized, and that the connecting principles may therefore need to take different forms. The Identity theory fails to account for some of the possibilities of the case, and is therefore an inadequate account of the logic of reduction in psychology.

2. *Model-Reduction*

I presented the necessary and sufficient conditions for explanatory reduction in psychology in terms of the relation of a psychological *model* to physiology. Will such a reduction therefore be an instance of model-reduction? It will be recalled that both the Identity theory *and* model-reduction claim that the connecting principles relating the two classes of terms will be identity statements (see pp. 32, 35). If this is so, what is the relation of model-reduction to derivation-reduction? If we can map a psychological model on to the nervous system, can we also logically derive psychological sentences from physiological ones?

A. *Models and Theories*

To make a start on these questions we need first to clarify the relations between describing a model and presenting a theory. Only then can we understand what distinction there is, if any, between deriving one *theory* from another and relating a *model* to a theory.

In Chapter I, I defined a 'model' as some system or phenomenon which is fairly well understood and which is thought to be analogous in some ways to the phenomena under investigation. An example is a system of tiny billiard-ball-like particles, employed as a model for a gas. Some of the features of the system of tiny particles are thought to be analogous to the features of a gas, and our understanding of the behaviour of a system of tiny particles in random motion helps us to understand some of the features of a gas. Is there any difference between describing this model for a gas and presenting a theory of gases? Philosophers of science have defined a rather precise relationship between models and theories, which requires a somewhat technical detour to explain.

According to Nagel, Braithwaite, and Carnap, a theory is a *partially interpreted calculus.*[2] Theories have syntactic and semantic aspects. Syntactically, a theory is a formal calculus. One adds to this calculus certain rules which partially specify the meaning of terms in the calculus. These rules are called 'correspondence rules'; they provide the calculus with a 'partial interpretation' because they specify a meaning for only *some* of its terms under *some* conditions. Only when some of the terms in the calculus are given a meaning by the correspondence rules do we have a theory.

A calculus is simply a set of symbols and rules for writing well-formed formulae with those symbols. The symbols fall into two classes: logical operators (typically set as 'and', 'or', 'not', 'if . . . then', 'if and only if', 'all', and 'some'), and non-logical operators. The non-logical operators are terms standing in for names, variables, and predicates of any order (ascribing states, properties, or relations). The calculus specifies rules for writing strings of these symbols as formulae, and for manipulating well-formed formulae to produce other well-formed formulae.

[2] For clarification on this see Rudolf Carnap, 'Theories as partially interpreted formal systems', in *Readings in the Philosophy of Science,* ed. Brody; Rudolf Carnap, *Foundations of Logic and Mathematics*, International Encyclopedia of Unified Science, Vol. I, no. 3. (Chicago: University of Chicago Press, 1939); R. B. Braithwaite, *Scientific Explanation* (Cambridge: Cambridge University Press, 1959), Chs. 2 and 3; Ernest Nagel, *The Structure of Science*, pp. 90-5.

To give the calculus a meaning one must assign a meaning to some of its name-terms and predicate-terms. This is the job of the correspondence rules.[3] They pick out a domain to which the calculus is to apply—the phenomena under investigation which the theory is meant to explain. The correspondence rules relate some of the non-logical terms in the calculus with certain previously understood observation terms which pick out individuals, states, or relations in the given domain. Some of the name-terms in the calculus are linked to observation names which pick out individuals in the domain. Some of the predicate-terms in the calculus are associated with observational predicates which denote certain states, properties, or relations in the domain. Finally, some of the variable-terms in the calculus are given a definite scope, so that they range over a particular collection of individuals in the domain.

These correspondence rules may take many forms. They may give an explicit definition for a predicate-term in the calculus by substituting an observational predicate for it.[4] They may give an observation term as a necessary condition for applying a theoretical term in the calculus, or they may specify a sufficient observational criterion for applying a theoretical term. They may link a given theoretical term to a particular observable response under a given test condition. In general the correspondence rules simply relate some of the name-terms, variable-terms, and predicate-terms in the calculus (the so-called 'theoretical terms') to previously understood observation terms applying to the intended domain.[5]

An example is urgently required. Suppose that we wish to

[3] For clarification see Rudolf Carnap, *An Introduction to the Philosophy of Science,* ed. Martin Gardner (New York: Basic Books, 1966), Ch. 23-8; Rudolf Carnap 'The methodological character of theoretical concepts', in *Minnesota Studies in the Philosophy of Science,* Vol. I, eds. Herbert Feigl and Grover Maxwell (Minneapolis: University of Minnesota Press, 1956); Carl G. Hempel. 'The theoretician's dilemma', in his *Aspects of Scientific Explanation,* especially pp. 187-9, 206-10; Braithwaite, *Scientific Explanation,* Ch. 3; Nagel, *The Structure of Science,* pp. 91-5, 97-105.

[4] The interpretation would thereby provide the theoretical term with both sense and reference. Interpretations which employ different substituends to pick out the referent of a given term will thereby give that term different senses. See Benson Mates, *Elementary Logic,* 2nd edn. (New York: Oxford University Press, 1972), pp. 76-7.

[5] See Hempel, 'The theoretician's dilemma', pp. 187-9, 206-10 for an account of some different forms of correspondence rules.

explain a generalization which holds true for gases of constant volume:

> For any gas *x* of constant volume, if the temperature of *x* increases then the pressure of *x* increases.

We might propose to explain this generalization by postulating some hypothetical state *M* which is an invariable consequence of the increase in temperature of a gas at constant volume, and some hypothetical state *F* which in gases of constant volume invariably results in increased pressure. More formally, we propose the following calculus. '*M*' and '*F*' are the only non-logical terms of the calculus (they are predicate-terms) and the following is a well-formed formula:

For every *x*, if *x* is *M* then *x* is *F*.

We then give the calculus a meaning by specifying that '*x*' is to range over gases of constant volume, and adding the correspondence rules:

> For any gas *x* of constant volume, if the temperature of *x* increases then *x* is *M*.
> For any gas *x* of constant volume, if *x* is *F* then the pressure of *x* increases.

Notice that these correspondence rules only partially specify the meaning of our terms '*M*' and '*F*'; one of them gives only a necessary condition for '*F*' in observation terms, while the other gives '*M*' only a sufficient condition. The calculus and correspondence rules together give a set of relational characterizations of the hypothetical terms '*M*' and '*F*', linking them to each other and to the observation terms 'increased temperature' and 'increased pressure'. Our theory proposes that there are states which are linked in these ways to each other and to those observable states, and it thereby explains the correlation between those observable states. Specifically, the calculus and correspondence rules allow one logically to derive the sentence:

> For any gas *x* of constant volume, if the temperature of *x* increases then the pressure of *x* increases.

Once the theoretical terms '*M*' and '*F*' are linked to observable consequences, one can derive empirical predictions from the theory.

One may wonder how a partial interpretation can suffice to give theoretical formulae empirical meaning. The answer is that the calculus itself specifies a set of relations (in terms of logical operators 'and', 'or', 'not', 'if . . . then', and 'if and only if') between any given theoretical term and other theoretical terms. Any particular predicate-term or name-term is given a precise syntactic place in the calculus by these relations to other theoretical terms. To tack down some of the theoretical terms to observables in this way suffices to connect the entire structure of theoretic terms to the domain, and give at least indirect meaning to every term in the calculus. Uninterpreted terms will be related precisely to terms which in turn are related to observation terms. The calculus specifies an abstract structure of relationships which are purportedly satisfied by the theoretical states and processes in the domain. In this way, if some of the theoretical terms can be attached to observational states or relations, the entire structure of relationships is attached to the domain.

The hypothetical states in the domain are characterized only in terms of their relations to one another (represented by the calculus) and to observables (as given in the correspondence rules). For example, what it means for a gas of constant volume to be in an *M* state is simply that it is in a state which is an invariable consequence of increased temperature and which leads invariably to an *F* state. The meaning of the theoretical term '*M*' is given by the following definition:[6]

> If there exists some state which in gases of constant volume is an invariable consequence of increased temperature and which in turn invariably results in *F* states, then that state is an *M* state.

Theoretical terms are given meaning solely in terms of their relations to one another and to observables. This is one of the

[6] Claims about the meaning of theoretical terms involve one in the problem of distinguishing sentences true by virtue of the meanings of their words from sentences true by virtue of their correspondence with empirical fact. This 'analytic-synthetic' distinction has been a major area of controversy in analytic philosophy in the past twenty-five years. For a defence of the position adopted here see Carnap, *Introduction to the Philosophy of Science*, pp. 265–74.

reasons why it was argued above (pp. 68–71) that the terms for internal variables in a psychological model specify their referents only in terms of their relations to other internal variables, stimulus variables, and response variables.

We can now elucidate the relation between models and theories. Along with correspondence rules there is a second way in which the non-logical terms in a calculus can be given a meaning. One provides all of those terms with a second interpretation, which assigns to the appropriate theoretical terms predicates and names from some other realm of discourse.[7] Those predicates and names describe some system which is relatively well understood and which is thought to provide an analogy for the processes and states postulated by the theory. This second interpretation is distinguished from the correspondence rules by two major features. The first is that unlike the correspondence rules, it assigns predicates to *every* predicate-term, and names to *every* name-term in the calculus. Second, the terms it assigns to the theoretical terms of the calculus are taken from a field of discourse distinct from that of the domain assigned by the correspondence rules.

For example, consider the calculus and correspondence rules given above to explain a generalization about gases. We can give this calculus a second interpretation as follows: consider a system of tiny particles in rapid motion, whose collisions with each other and with the walls of their container are elastic. The terms we use to describe such a system are certainly different from those we use in investigating the properties of gases. But we can assign terms describing some features of this system of tiny particles to the theoretical terms '*M*' and '*F*', as follows. To the predicate-term '*M*' we assign the predicate 'increase of the mean kinetic energy of the particles', and to the predicate-term '*F*' we assign the predicate 'increase in the force exerted by the particles on the walls of their container'.

A *model* as defined in Chapter I is simply the system or phenomenon described by the second interpretation assigned

[7] See Nagel, *The Structure of Science*, pp. 94–7, 107–17; Braithwaite, *Scientific Explanation*, Ch. 4; R. B. Braithwaite, 'Models in the empirical sciences', in *Readings in the Philosophy of Science*, ed. Brody; Hempel, *Aspects of Scientific Explanation*, pp. 433–47.

to a calculus. This system is thought to be analogous in some ways to phenomena in the domain of the theory, and that is why it is chosen as a model for the calculus. The term 'model' is also applied in some contexts to the second interpretation itself, and to distinguish these two senses of the word when necessary I shall call one a 'model-system' and the other a 'model-interpretation'.

One consequence of giving a calculus a second interpretation is that the calculus no longer presents merely an abstract characterization of the structure of relations satisfied by one's hypothetical states and processes. Instead the theoretical terms seem to describe some actual states or processes of some actual system. Our formula

For every x, if x is M then x is F

is given only partial meaning by the correspondence rules, since those rules do not specify the meanings of '*M*' or '*F*' for gases whose volume changes, or whose temperature or pressure either remains constant or decreases. However, if we accept the model-interpretation for '*M*' and '*F*', then we can read the theoretical formula as follows:

> For every gas x at constant volume, if the mean kinetic energy of the particles of x increases, then the force exerted by the particles of x on the walls of their container increases.

The calculus conjoined to the second interpretation furnishes *descriptions* of a model-system.[8]

What is the difference between describing a model-system (presenting a calculus and a model-interpretation for it) and presenting a theory (a calculus and a set of correspondence rules)? According to proponents of derivation-reduction, the main difference is that model discourse is by definition suppositional, so that while the model-system can provide analogies for thinking about the real system, one is not committed to existence claims on behalf of the model. Model-systems are heuristic devices, suggestive or illuminating analogies, suppositions which can open new avenues of research and suggest revisions in theories. But the calculus and its correspondence rules remain central to reduction; at best a model-system

[8] See Nagel, *The Structure of Science*, pp. 94–5.

provides a set of analogies for thinking about the abstract structure of relationships postulated by the theory.

Proponents of model-reduction claim that models are the central medium of expression for theoretical activity, and that formal calculi are merely aids to thinking about one's model.[9] On this view, the central condition for reduction is to find a model-system *M* in one discipline for a theory T' in another. One enters existence claims for the model and on that basis explains a theory in some other discipline.

B. *Virtues of Model-Reduction*

The reduction of psychological models to physiology does prima facie show some of the features of model-reduction. The first and most obvious is that psychological theorists rarely employ partially interpreted theoretical calculi, nor do they often make explicit logical derivations of observations or predictions from theoretical postulates. Rarely does one find all of the assumptions and proposed general laws of a hypothesis explicitly listed, and unstated assumptions, restrictions upon generalizations, and escape clauses are often found in derivations of predictions from theories. Such derivations are rarely *logical* derivations. It is much more natural to describe theorizing in psychology as a matter of describing hypothetical mechanisms which could produce the observed stimulus-response relationships. Theoreticians do not employ partially interpreted terms; instead they seem to *describe* straightforwardly certain internal parts and their functional relationships. They give a functional characterization of a mechanism which would manifest the observed stimulus-response relationships. I suggest that in describing such a mechanism, one is describing a *model-system* for the behavioural phenomena in question.

For example, consider the mechanism described by J. A. Deutsch in his book *The Structural Basis of Behaviour.*[10] (See Figure 4.1.) Deutsch names the parts of this machine with terms such as 'link', 'analyser', and 'receptoral system',

[9] For arguments along these lines see Harré, *The Principles of Scientific Thinking*, pp. 3–21, 35.

[10] J. A. Deutsch, *The Structural Basis of Behaviour* (Cambridge: Cambridge University Press, 1964).

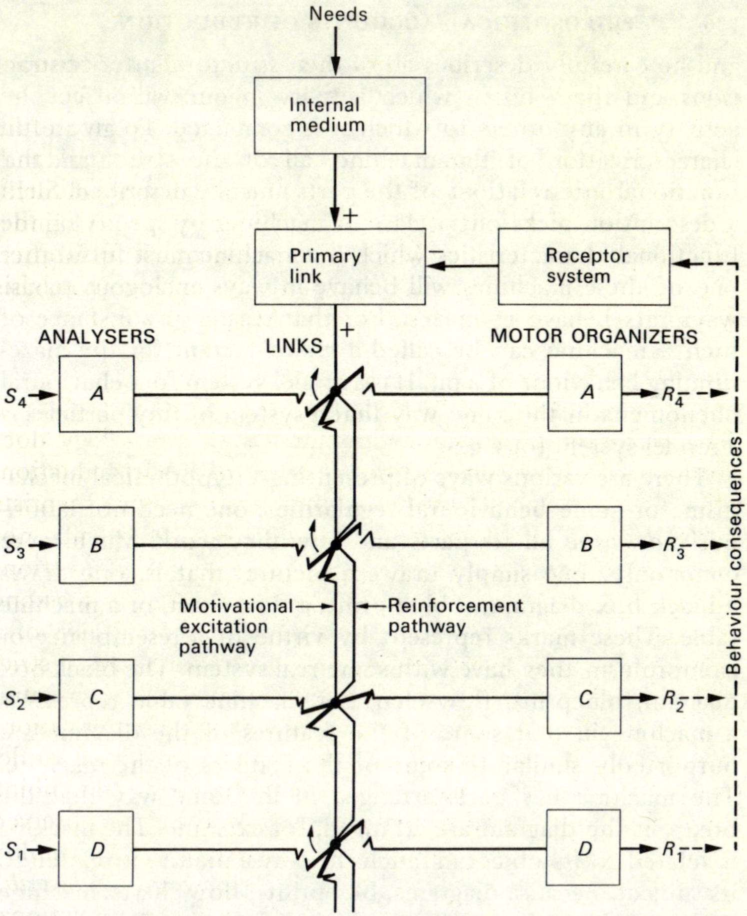

Fig. 4.1. Blueprint for Deutsch's maze runner. An analyser is activated by the occurrence of the appropriate stimulus, closing the link. Excitation proceeds down the motivational pathway from the primary link until it reaches a closed link, where it proceeds to a motor organizer and prompts a particular overt response. The system is shown with analyser *C* firing, closing that link and sending the motivational energy to motor organizer *C*. Learning occurs when owing to a succession of stimuli and responses, sets of analysers, links, and motor organizers are arranged in an ordinal progression. The link closest to the primary link is related to the stimuli and responses in the goal box. Stimuli arising from a consummatory response activate the receptor system, which inhibits the primary link, shutting off the system. (Based on J. A. Deutsch, *The Structural Basis of Behaviour* (Chicago: Chicago University Press, 1960), pp. 25, 40, 41. Copyright © 1960 by University of Chicago Press and reprinted with their permission.).

and he carefully describes all of their structural inter-connections and the ways in which activity in one will affect the activity in any others to which it is connected. To give a full characterization of the machine, all of the structural and functional interrelations of the parts must be described. Such a description picks out a class of machines by specifying the functional characteristics which the machine must fulfil. Any one of these machines will behave in ways analogous to the ways rats behave in mazes. For that reason any instance of such a machine can be called a *model-system* for the maze-running behaviour of a rat. It is a model-system for behavioural phenomena in the same way that a system of tiny particles is a model-system for a gas.

There are various ways of presenting a hypothetical mechanism for some behavioural regularities; one need not laboriously *describe* all its parts and how they work. Much more commonly, one simply draws a picture; that is, one draws a black box diagram, a blue print, a flow chart, or a machine table. These marks represent by virtue of a resemblance or isomorphism they have with some real system. The black box diagram, blueprint, flow chart, or machine table represents a machine in that some of the features of the diagram are purportedly similar to some of the features of the machine. The machine has parts arranged in the same way that the boxes in the diagram are arranged, for example. The marking is related to its object in much the way a picture is related to its subject. Because diagrams, blueprints, flow charts, machine tables, and graphs employ a pictorial (or iconic) mode of representation, I shall call them all *icons*.

Sometimes when we say that a theoretician presents a model for some phenomenon, what we mean is that he or she presents a diagram, chart, blueprint, or picture which represents hypothetical processes producing that phenomenon. One of the many senses of the word 'model' is therefore that of *icon*.[11] This sense of the word is related to that of 'model-system', for in effect icons are pencil-and-paper models for the phenomena in question.

In accord with the model-reduction account, icons make

[11] Harré claims that models—in the sense of *icons*—are usually employed as the 'vehicles of thought' in theoretical activity.

use of analogies to link them to the real system under investigation. The mere collection of lines on paper does not suffice to pick out a class of mechanisms, for one must also give some characterization of the jobs of the parts. To do this one labels parts of the icon with terms drawn from some other realm of discourse which provide analogies for the functions of the various parts. In effect one provides the icon with an interpretation: an assignment of terms from other realms of discourse for all its parts. For example, one employs terms from servomechanism theory ('comparator', 'negative feedback', 'transducer'), automaton theory ('and-gates', 'sub-routine'), computer technology ('encoder', 'memory store'), electronics ('filter', 'gain'), communication technology ('channel', 'noise', 'signal detection'), and finally, from ordinary psychological discourse ('decision mechanism', 'feature detector', 'arousal mechanism'). Variables in the diagram can also receive names from ordinary psychological discourse, such as: expectancy, fear, frustration, hope, safety signal, relief, reward, anxiety.

One job of such terms is to specify the functions of the parts pictured by the icon. The icon with its interpretation (labels for parts and variables) suffices to pick out a class of mechanisms, any one of which is a model-system for the behaviour in question. The mechanism it specifies would have the same input-output relationships as the real system, and purportedly would have internal processes similar to those of the real system.

A second job of the terms labelling the diagram is to provide some initial analogies relating the diagram to the real system. In a pencil-and-paper model of perception and attention, one might predict certain input-output relations by employing the term 'filter' to characterize the function of one of the parts. One might propose for example that attentional mechanisms behave as if they were filters for sensory input. This analogy suggests certain experiments. Furthermore, designing hypothetical mechanisms (model-systems) for behaviour can often suggest explanations for previously unexplained phenomena. For example, Deutsch's machine will show insight behaviour whenever a row of links and analysers is short-circuited by presentation of a stimulus out of the normal order. We can give a tentative explanation of insight behaviour in the rat:

it is as if a row of links and analysers controlling motor sequences were short-circuited. This explanation removes some of the sense of mystery around the term 'insight' and provides a way of understanding what is going on. Hypothetical mechanisms can organize and classify bits of unrelated data, suggest new stimulus-response relationships to investigate, and in general act as an extremely useful tool in empirical inquiry.

If one's description of a model-system is sufficiently specific, one can artificially construct a machine satisfying the description, and test its behaviour in various situations. Such an artificially constructed model-system is called a *simulation*. On the basis of his description of a hypothetical mechanism, J. A. Deutsch proceeded to construct a maze runner which could learn how to run mazes; and he claimed his machine would perform in a maze, and improve its performance, in the same way as a rat. Furthermore, Deutsch was not content with the claim that his maze runner was a useful device for stimulating inquiry and suggesting new hypotheses. He also claimed that the processes inside the machine which enabled it to learn were similar to those inside a rat.

In so far as a simulation is a good one, it will purportedly satisfy the same functional description as the nervous system of the animal in question: each part of the simulation will be analogous to some part of the nervous system, and it will do the same job as that part. In this way some of the functional characteristics of the device will be identical to some of those of the nervous system. Far from providing mere analogies for neural functioning, the model-system represents some of its actual features.

If reduction is successful, particular parts of the nervous system are identified as parts of one's hypothetical mechanism, with the same interrelations and the same jobs. The claim is not merely that the organism acts *as if* it contained a negative feedback system, or a kind of selective filter, or a mechanism for adjusting the gain on inputs to other parts—but that there actually *is* a negative feedback system, a filter, or a gain adjuster in the nervous system. (The analysis in Section 4 of Chapter II was intended to show how one establishes that a particular neural structure carries out one of these jobs.) Terms such as 'feedback', 'encoder', 'comparator', or 'channel' drop their role as mere analogical tools—ways of thinking of the workings of the black-box—and emerge as accurate descriptive terms, characterizing the functional role of particular neural arrangements. The connecting principles relating the two sets of terms are identity claims, identifying the nervous system as an instance of the same sort of mechanism as one's model-system.

The final condition in model-reduction is that one *explain* the connections between one's model-system and the phenomena in the domain of the reduced theory. To do this one shows that features of the model-system are exactly analogous to the features of the phenomena in the domain of the reduced theory (see above, pp. 35–6). If one can successfully map a psychological model on to the nervous system, then one can explain the connecting principles which relate a psychological term to its neurophysiological correlate. Specifically, to explain why a given neural state is a value of a particular psychological variable, one shows how the neural state has a functional role isomorphic to that assigned to the variable in the model. For example, one can explain why a given neuron is a reward neuron by showing how, given the connections of that neuron with other parts of the system, the firing of that neuron increases the probability that an immediately preceding response will be repeated. To explain why a given structure mediates arousal, one shows how, given the relations of that structure to other parts of the nervous system, its activity leads to increased intensity of response, greater efficiency of response, and greater efficiency of learning.

C. *Some Problems with Models*

The difficult question remains of whether derivation-reduction *cannot* account for those features of explanatory reduction in psychology which are consistent with a model-reduction account. On a derivation-reduction view, one can accept that models play a useful and important role in research, while still insisting that explanatory reduction is a relation between the theories which are the end-product of such research. Nagel, Braithwaite, and Hempel can all accept that it may be easier to think in terms of a familiar model-system rather than try to derive predictions from a partially-interpreted calculus. Thinking in terms of a model may be very useful in suggesting new lines of research, new laws to account for data, and revisions in one's theory to make it more precise.[12] A model-system for a theory may be useful in reduction by suggesting connections between that theory and others. For example, thinking of a gas as a system of tiny particles suggests the link between 'pressure' and the force exerted by those particles on the walls of their container.[13] Such accounts are consistent with the idea that model-systems are often the 'vehicles of thought' in the sense that they suggest experimental tests and serve to collect and organize data. In all of these ways model-systems play a vital role in inquiry.

However, on these views model discourse must remain suppositional. In proposing a model one is in effect saying 'Suppose it worked this way . . .' or 'It is as if it works that way . . .'. These suppositions or analogies can often cast light on murky issues, suggest new hypotheses and new lines of research, and contribute mightily to empirical explorations. But to employ models in explanations or enter existence claims for some features of models is to misunderstand their role.

Consider the claim that describing a hypothetical mechanism can in some sense explain observations. Deutsch's machine shows insight because the firing of an analyser sometimes short-circuits a row of links. When we consider the insightful behaviour of a rat, therefore, the machine suggests that the rat behaves as it if had a row of links which is short-circuited by a stimulus. This supposition can admittedly cast light on the subject and open new avenues of investigation. But does the supposition *explain* the behaviour of the rat? In saying 'It is as if the rat has a row of links which are short-circuited', we are not claiming that the rat *does* have a row of links in its nervous system. As long as we employ 'as if' or 'hypothetical mechanism' discourse—which is a very useful form of discourse—we do not commit ourselves to the existence of the mechanisms described. Our model-system may provide a good analogy for the phenomena in question, but that is all.

Referring to a short-circuited row of links can explain the

[12] Nagel, *The Structure of Science*, pp. 112–13.
[13] Ibid., p. 133.

insightful behaviour of the rat only if we accept that the machine does not provide merely an analogy for the processes involved in maze learning, but claim that a mechanism functionally similar to a row of links exists in the rat's nervous system. When we do this, we add to our theory. To use the model-system to explain data is to make claims about the real system; at the very least one assumes that it is like the model-system in the relevant respects.

Proofs in geometry are sometimes invalid if the geometer tacitly assumes that the points, lines, and planes mentioned in his axiom system share the features of the ordinary points, lines, and planes described by an interpretation for the terms in the axiom system. Similarly, to explain heat conduction in gases by referring to transfers of kinetic energy in collections of tiny particles is to make some tacit assumptions about molecules that otherwise may not be part of one's theory: that their collisions are elastic and that no cohesive forces act between them. In so far as one relies upon analogies between the model-system and the real system to explain observable regularities, one tacitly assumes that the entities of the real system share the relevant features of the model-system.

This same sort of problem is involved in the claim that in reduction one enters existence claims for the mechanisms described in one's model-system, or that one identifies phenomena in the domain of the reduced theory as an instance of the same sort of system as the model-system. The question which the classical view would prompt here is: if we enter existence claims for the mechanisms of the model, in what sense is it still a *model*? Isn't the change here precisely one of absorbing parts of the model into the theory?

In effect the disagreement is over the semantic question of whether or not it is accurate to say that theoretical reasonings still employ *models* after existence claims for those models are made. Carnap, Nagel, and Braithwaite would say that once existence claims are entered on behalf of the model, those claims become part of one's *theory*, and one has abandoned model discourse. Whereas on a model-reduction account one would presumably say that there are other features of the scientific reasonings which justify the claim that model discourse is still employed.

The question may be a purely semantic one, for under some conditions the requirements for derivation-reduction and model-reduction are equivalent. It is easy to show that once existence claims are entered on behalf of one's model, a description of a model-system is equivalent to a theory. We can describe a black box model (the one on page 48 for example) with sentences of the following form:

There is a part F_1 with inputs s_1 and s_2 and output x_1 such that $x_1 = f_1(s_1, s_2)$

•

•

There is a part F_5 with inputs x_1, x_2 and output x_5 such that $x_5 = f_5(x_1, x_2)$

•

•

There is a part F_{11} with input x_7 and output r_4 such that $r_4 = f_{11}(x_7)$.

This collection of sentences is a theory. If we can reduce the model by relating it via general identities to the nervous system, then we can derive the above theory from a set of sentences about the nervous system and those general identities. Suppose we have general identity statements which show that our model is isomorphic to the nervous system, so that there is some neural structure S_1 which does the job of part F_1 in the model, there is some neural variable u_1 which is isomorphic to the input variable s_1 in the model, and so on. To show that the nervous system is an instance of the same functionally-characterized machine as the model system, we would need to establish that:

There is a structure S_1 with inputs u_1 and u_2 and output v_1 such that $v_1 = f_1(u_1, u_2)$

•

•

There is a structure S_5 with inputs v_1, v_2 and output v_5 such that $v_5 = f_5(v_1, v_2)$

•

•

There is a structure S_{11} with input v_7 and output w_4 such that $w_4 = f_{11}(v_7)$.

But now from the above sentences about the nervous system, and the general identities relating S_1 to F_1, u_1 to s_1, v_1 to x_1, and so on, we can *derive* the above sentences comprising our theory. So in any case where we can achieve a model-reduction with general identities, we can construct an equivalent derivation-reduction, using those general identities as connecting principles. In this case, there is no essential difference between model-reduction and derivation-reduction, for whenever one can be achieved, so can the other.

It is easy to see why those conditions on reduction are equivalent. A model-system is some familiar system whose states and processes are related to each other in just the way the hypothetical states and processes of the real system are purportedly related. If the model is to provide a useful tool for thinking about one's theory, a description of the model-system must be formally isomorphic to the psychological theory.[14] But now suppose we also show that the nervous system is isomorphic to our model, and we use identity claims to link the theory and the model to the nervous system. The psychological sentences, the description of the model-system, and the physiological sentences are isomorphic to one another, and if identity claims link them, a logical derivation of the psychological theory from physiology is a simple matter. It can be proved as a general theorem that if we identify the elements of a model with those of some system, and reduce the model by showing it isomorphic to the given system, then we can construct a derivation-reduction between the sets of sentences describing the model and the system.[15]

One response to this argument is to point out that perhaps the psychological model will not be connected to physiological terms with general identity statements. Such identity statements are crucial, for mere formal isomorphism between sets of sentences is insufficient for a logical derivation.[16] However, if we need to use disjunctive, non-specific, or incomplete

[14] On formal isomorphism see above, pp. 83–4. See also Hempel, 'Aspects of scientific explanation', pp. 435–6.

[15] See Schaffner, 'Approaches to reduction'. Schaffner shows that a Nagel-Woodger-Quine reduction is sufficient to construct a Suppes reduction. To construct the former from the latter, one must add the condition that the domains of the models in the Suppes reduction are identical.

[16] See Sklar, 'Types of inter-theoretic reduction', pp. 109–24.

connecting principles, we could not derive the psychological theory from physiology, although we could show that the psychological model is functionally isomorphic to the nervous system (I argued this above, pp. 125-7). In that case we would have a genuine non-equivalence of derivation-reduction and model-reduction, for the latter could be achieved even though the former could not. If functions are loosely localized, it may therefore be clearer to give a model-reduction account of the conditions on reduction than a derivation-reduction account. But this potential non-equivalence of the two accounts is perhaps not very significant, for there is little evidence currently obliging us to accept the hypothesis of loose localization of function.

A second response to the claim that model- and derivation-reduction are equivalent is that even after theory changes by absorbing parts of the model, scientific reasoning will continue to show features of model discourse. Perhaps we shall still not have precisely-formulated theoretical postulates or strict logical derivations. One may still employ analogies and icons in one's explanations of behaviour. The relation between experimental findings and the models they support or infirm may not be that of strict logical derivation or contradiction. For all these reasons, strict derivationist accounts are inaccurate descriptions of theoretical activity in science, and the use of models is the norm.

It is certainly true that the use of models allows one to carry on theoretical activity which does not show precisely-formulated assumptions or strict derivations of predictions from theories. But I suggest that this is because such precision and logical sophistication is not needed—and is perhaps a hindrance—in early exploratory stages of inquiry. Model discourse is simpler and more fertile and suggestive than formalized theory, and so when inquiry is chiefly concerned with gathering data and discovering some of the basic empirical regularities in a field, use of models may be more profitable than use of uninterpreted calculi and logical derivations. Even though one finds considerable fuzziness about the precise postulates and existence claims asserted in order to explain the data, nevertheless such woefully imperfect theoretical discourse is sufficient at this stage to allow choices to be made between rival models on the basis of experimental evidence.

One can admit all this and still insist that the absence of explicitly-formulated postulates and strict derivations, and the use of unstated assumptions or escape clauses results in a lack of clarity and precision, and that as the discipline advances and theory becomes more precise, these features will gradually vanish. Certainly in the more advanced sciences, theories and predictions show more of the features ascribed to them by the derivationist accounts. As models become more refined, the logical machinery in one's theories must be made more precise, and the implicit assumptions formalized. This has certainly happened already in psychophysics and branches of mathematical psychology, where the reasonings are relatively strict and the theories relatively explicit.

In summary, both model-reduction and derivation-reduction accounts can accept that models are crucial tools of research, illuminating experimental findings and suggesting new experiments; that the model is originally linked to the real system by relations of analogy, but that some features of the model may be recognized as genuinely similar to those of the real system; that reduction is a matter of entering existence claims for the model-system, and showing that the real system satisfies the same functional description as the model-system. The two accounts differ over the question of whether models have uses other than those of suggestive analogies, and in particular whether it is accurate to say that theoretical discourse employs models even after existence claims have been entered for them. This question is largely a semantic one, for if general identities link the model to the nervous system, the conditions for model-reduction are equivalent to those of derivation-reduction. It may, however, be clearer to phrase the conditions on reduction in terms of models in cases where general identities are not found or where research is in its infancy and precise formalization and strict derivation of predictions are not needed for experimental progress.

3. *Four Functionalist Anti-Reductionist Arguments*

Some philosophers have used the claim that the connecting principles between psychological models and neurophysiological mechanisms will be statements of functional isomorphism to support the claim that such models cannot be reduced to neural mechanisms. The final task of this chapter is to show why these arguments do not hold. My general point is that even though the connecting principles may be logically odd, they can still be used to achieve a reduction.

A. *Identity of Kinds*

One oft-cited feature of the connecting principles relating a functional characterization of a system to its internal structures is that different neural states may at different times play a given functional role. The mapping principles may need to associate a given psychological variable with a disjunctive list of neural states (ϕ_1 or ϕ_2 or . . .) which are its values at different times.

This is thought by some philosophers to provide grounds for believing that reduction in psychology is not possible. Apparently they draw this conclusion because they believe that the need for the above disjunctive list renders a derivation of the psychological laws impossible. For example, Fodor claims that what he calls 'identity of kinds' is a necessary condition for explanatory reduction; that unless one finds a particular neurophysiological term which picks out just those states picked out by the psychological term, the reduction cannot work.[17] By 'identity of kinds' he means that for every psychological term there is some neurophysiological term which picks out exactly the same class of states as the psychological term. If this is a requirement for reduction, and different instances of the same psychological state are instances of different neural states, then the reduction of psychological models to neurophysiology is precluded.

My first objection to this argument is that is has not been established that the connecting principles will or must take the form of disjunctive lists, but merely that they *may* take that form. The argument ignores the plausible alternative that localization of function will be specific and of general validity.

Furthermore, consider the claim that a necessary condition for deriving the laws of a reduced science from those of a

[17] Fodor, *The Language of Thought*, pp. 11-13.

reducing science is that one finds terms in the reducing science coextensive with each term in the reduced science. Although initially it seems plausible, it is easy to see that such identity of kinds is *not* a logical requirement for derivation. Suppose 'if *P* then *Q*' is the law of our reducing science, and 'if *X* then *Y*' is the law of the reduced science we wish to derive from it. It is a matter of elementary logic to verify that 'if *X* then *P*' and 'if *Q* then *Y*', conjoined to the former sentence as the connecting principles, will suffice to derive the latter sentence. One does *not* need to establish that '*X*' and '*P*', and '*Y*' and '*Q*', are coextensive, in order for the derivation to succeed.

The mere disjunctive form of the connecting principles does not preclude a derivation, as long as we can pick out just those neural states and structures doing a job at different times. As long as particular identities at any time can be established, the possibility that those particular identities pick out different states at different times does not matter.

This purely logical point can be supplemented by some counter-examples to Fodor's claim. In fact, in *most* cases of micro-reduction, different numerical instances of the macrostate identified in the reduced science are instances of qualitatively different microstates identified in the reducing science. Consider the standard example of the reduction of thermodynamics to statistical mechanics. One can have two samples of gas at the same temperature, pressure, and volume, although one is fluorine gas and the other is oxygen. Although we have two instances of gas at a certain temperature, each is an instance of a different microstate, since the molecules in one are fluorine molecules, while those in the other are oxygen. Yet in both cases one can reduce explanations of the relations of the temperature, pressure, and volume of the gas to statistical mechanical explanations.

One's level of analysis determines what counts as similarity or difference between states. Features at the macrolevel may be sufficient to back a claim that we have two instances of the same state, yet at a microlevel those instances will display many differences. In one sense, all human faces are the same, and yet in another sense every human face is different from every other one. Similarly, we can say that two houses have the same structure even though one is built of sandstone brick and the other of identically-proportioned limestone brick.

This analysis is intended to show how unimportant for the logic of reduction is the fact that one may need to use a disjunction of different neurophysiological predicates to pick out the values of a given psychological variable. It is a general feature of micro-reductions that different instances of the macrostate are instances of different microstates. Identity of kinds is far too strong a condition even for derivation-reduction. Indeed, I argued above that reduction can be achieved with disjunctive, incomplete, and non-specific connecting principles. Such connecting principles are far weaker than those establishing identity of kinds, yet they may suffice to establish a model-reduction.

B. *Problems of Scope*

A second feature of psychological and physiological characterizations of a system is that a psychological model may have wider scope than the descriptions of the physiological mechanisms which instantiate it. The model may be applicable to two species even though the neural systems instantiating the model are different in the two species. This seems to give grounds for an anti-reductionist argument which proceeds as follows: if the psychological model is of greater scope than the neurophysiological descriptions given by the mapping principles, then one cannot derive the psychological laws from the neurophysiological laws. One cannot give a neurophysiological explanation of the psychological phenomena, since the same psychological phenomena are manifested in different neural structures, and the psychological laws have greater explanatory power than the physiological ones. Therefore psychological models cannot be reduced to neurophysiology.

Notice that this argument relies on the tacit assumption that reduction, if it is to occur at all, must be an instance of derivation-reduction. It is only this assumption which licenses the inference from the supposed greater generality of the psychological model to the claim that it cannot be reduced to physiology. Other types of reduction are still open, however.

Furthermore, it does not even follow from the greater generality of the psychological model that it cannot be given a neurophysiological explanation. All that follows from it is that the neurophysiological explanation of the psychological laws will not be as general as those laws themselves. For each species to which the model applies, however, one could give such a neurophysiological explanation. That is, within the scope of a collection of mapping principles, one can explain why a particular psychological function is associated with a particular structure by showing how the structure could play the appropriate role in the work of the whole. We may need two sets of mapping principles to map a model of classical conditioning on to both rats and sticklebacks, yet once we have those principles we can certainly give neurophysiological explanations, for both species, of their respective learning capacities. The explanations will be identical in form, and differ only in the structures they pick out.

C. *Valve Openers and Camshafts*

Fodor advances another argument against reduction based on a functional account of psychological processes. He points out that 'the relation between functional analysis and mechanical analysis is not at all like the relation between macroanalysis and microanalysis'.[18] Since functions are abstract characterizations of certain system processes, they do not have parts, and one makes a conceptual mistake if one tries to give a microanalysis of them. Furthermore, since psychological terms are names for functional states, one cannot give a microanalysis of a psychological state, and so one cannot identify a psychological state with a neural state. 'Psychological states are not available for microanalysis'[19] and therefore cannot be identified with neural mechanisms. He concludes:

> It should be noticed that the view of the relation between psychological and neurological theories espoused here is to be distinguished from all varieties of reductionism . . . to attempt to reduce a functional account to a mechanistic account would be patently absurd.[20]

Fodor backs his argument with an example. In the sense

[18] Fodor, 'Functional explanation in psychology', p. 236.
[19] Ibid., p. 238.
[20] Ibid., p. 236.

that 'valve opener' names a particular function within an internal combustion engine, a valve opener does not have any components, and so a valve opener does not *consist of* a camshaft (even though the camshaft may be fulfilling that function). This is a necessary truth, according to Fodor. Similarly, it is also a necessary truth that a drive does not consist of a neural state. Functions do not have components. He says:

> There is no sense to the question 'What does a valve opener consist of?' where this is understood as a request for a microanalysis. Functions do not have parts; valve openers are not made of rods, springs, and atoms in the sense that camshafts are.[21]

The problem with this argument is that even though functions do not have components, this does not imply that functional states are not physical states, nor does it imply that the question of which particular physical states are instances of functional states is a meaningless one. Even the example is ambiguous on this score. One can use the word 'valve opener' to refer either to a function or to a class of components fulfilling a given function. It is true that functions do not have parts, but this does not imply that particular instances of valve openers are *not* instances of particular parts. To say that 'A valve opener does not consist of a camshaft' may mean several things:

(1) Valve openers need not be camshafts but can take different forms, even though particular instances of valve openers may be camshafts.
(2) A camshaft is a valve opener only if it is connected in the proper ways to other parts of the machine.
(3) 'Valve opener' is not a term denoting particular physical or structural attributes ('camshaft'), but is a functional term, describing the role the part plays in the work of the system.

One *can* appropriately ask about the components of a valve opener, that is, of a particular instance of a valve opener, a part fulfilling that function. In a similar way, one can ask about the neural states fulfilling certain functional roles within a psychological model.

[21] Ibid., p. 237.

D. Ad Hoc *and Incomplete Connecting Principles*

Putnam gives another argument against reduction, based on the character of the connecting principles involved.[22] He claims that explanatory reduction in psychology cannot occur because the connecting principles relating psychological to physiological terms would be statements of functional isomorphism (as described in Chapter II); therefore, he says, there could not be sufficient empirical evidence to justify accepting such principles, and their acceptance into theory would be *ad hoc* and would fail to simplify theory. Because psychological states are functional states, Putnam emphasizes, they may be instantiated at different times by different neural states. The statements picking out the neural states which at different times are instances of a given psychological state may be enormously complicated, and the simplest theory may be one that uses only psychological concepts and does not attempt to establish their physiological basis. There is a possibility that functions are so loosely localized that discovery of all the structures subserving a given function at different times is more or less impossible, leaving the connecting principles incomplete, and thereby precluding a derivation-reduction (see pp. 94–5).

Notice one way Putnam's argument could be strengthened. Many of the arguments about the impossibility of explanatory reduction in psychology could be converted into arguments that the various non-formal conditions on reduction will not be fulfilled. Perhaps the arguments about the logical form of our psychological concepts reveal difficulties in the attempt to formulate the necessary connecting principles between those concepts and the neural terms. Because the terms have such disparate logical features, the connecting principles would be in practice impossibile to formulate, too cumbersome, and *ad hoc*. They would fail to simplify theory. The common argument that an action is not a bodily movement, and that therefore explaining a movement does not explain an action,[23] could be reformulated as follows: 'The connecting

[22] Hilary Putnam, 'Philosophy and our mental life', in his *Mind, Language and Reality*, Philosophical Papers, Vol. II, pp. 291–303.

[23] See A. I. Melden, *Free Action* (London: Routledge & Kegan Paul, 1961), pp. 72, 171–84.

principles linking a particular action term (such as "making tea") to terms describing physiological events (such-and-such motor units firing, for example) would be enormously complicated if not in fact impossible to formulate. Even if one could find such connecting principles they would not simplify theory. Therefore explanations of actions cannot be reduced to explanations citing physiological happenings.'

Notice that this argument is not *a priori* or conclusive. It does not assert that the formal conditions on reduction will not be met; it attacks the possibility of fulfilling the non-formal conditions. I agree that if one could establish that the connecting principles relating psychological to physiological terms would be *ad hoc* and unverifiable, and would fail to simplify theory, it would follow that explanatory reduction in psychology is not feasible.

However, I do not think that Putnam has established that the connecting principles will have those features. More strongly, at least within the field of motivation and learning, some of the examples I have already considered show that at least some of the connecting principles are not as Putnam claims. Although Putnam's claim may apply to the connecting principles relating a term in ordinary language (such as 'making tea') to underlying physiology, when we turn to developed parts of the experimental discipline, his claim does not hold up.

First, hypotheses of the physiological structures underlying motivational functions have empirical content and are not *ad hoc*. Their empirical content is to be found in the mass of experimental work concerning localization of functions within the nervous system. The procedures for investigating which neural structures do the jobs assigned to boxes in the model were analysed in Section 4 of the second chapter (p. 73). When functions are localized, one gains evidence of the neural states corresponding to a particular psychological variable. Furthermore, there is little evidence obliging us to accept the pessimistic predictions of loose localization of function.

Secondly, the discovery of such connecting principles serves to simplify theory and advance knowledge. I have provided several examples which show how investigations of the underlying physiology can play an important role in choices between psychological models (see pp. 76, 113). Some of these investigations have been highly suggestive to the psychologist interested in the mechanisms involved in the organization of behaviour. If physiological psychology is aiming at a unified model of the functioning of the nervous system, then connecting principles giving neural values of the psychological variable are a valuable step towards the simpler theory.

In short, Putnam's characterization of the connecting principles relating psychological to physiological terms does not generally hold true. It may hold for terms in ordinary language, but perhaps this is because no developed models use such terms, and because investigators do not use them in their theorizing.

V

Eliminative-Reduction

The distinction between reduction and *replacement* is not always clearly made in the various functionalist arguments against the possibility of explanatory reduction in psychology. Many of the arguments show that because of the particular logical and semantic features of functional characterizations, such terms have explanatory and descriptive uses which cannot easily be fulfilled by terms naming structural characteristics. We can summarize Fodor's camshaft/valve opener argument by saying that when we call a part a valve opener, we are characterizing that part in a different way from when we describe its structural make-up, that is, characterize it as a camshaft. That the two concepts play different roles in contexts of description and explanation is taken to license the claim that explanatory reduction is precluded. Likewise, the problems raised by the purported requirement of identity of kinds between concepts in the reduced and reducing sciences, and by the fact that those concepts belong to different taxonomies, show that the psychological terms have uses which cannot easily be fulfilled by physiological terms. But if we assume that these arguments show that explanatory reduction cannot occur, we are confusing explanatory reduction with the replacement of explanations provided by one theory with those provided by another.

However, the assumption that explanatory reduction suffices to establish a replaceability of explanations is one accepted by Eliminative materialists, who claim that the reducing theory could entirely replace the reduced one in contexts of description and explanation. In Chapter I, I labelled the assumption that explanatory reduction establishes a replaceability of explanations the 'Replacement' assumption, and I showed how many of the arguments against the possibility of reductionism in psychology tacitly rely upon it. Will the Replacement assumption hold for the reduction of psychological models to physiology? The Replacement assumption describes just one of the conditions of eliminative-reduction, and so we must also ask: which other features of eliminative-reduction will be found in reductionism in psychology?

1. *Replaceability of Psychological Terms and Explanations*

A. *Logical Derivation and Replacement of Explanations*

To clarify these questions we must first understand two different senses in which one explanatory system can replace another. The word 'explanation', like the word 'proof', has different senses.[1] To explain something means (roughly) to render some puzzling phenomenon clear or intelligible, and the word 'explanation' can refer to the act of explaining something or to the sentences employed while explaining something. In either case an explanation is defined in relation to a question raised by a particular person (with certain beliefs and reasoning abilities) in a particular context, and hence what may be an explanation for one person may be unintelligible or unenlightening for someone else. If we use the word 'explanation' in this relativized, contextual sense, we are focusing on what I shall call the *pragmatic* aspect of explanation. Other aspects of explanation can be analysed. Abstracting from the particular contexts at hand and the particular people involved (with their beliefs, reasoning capacities, and so on) one can consider the formal relations between the sentences employed, and ask in general what is the relation between the sentences explained and those used to do the explaining. It is to this objective (non-contextual) or *formal* aspect of explanation that the derivationist account is meant to apply. That account claims that this objective sense of explanation corresponds with showing that certain sentences *logically imply* some other sentences. The conditions defining an adequate explanation in this sense of the word are independent of

[1] This distinction was proposed by Carl Hempel. See his 'Aspects of scientific explanation', pp. 425–8.

particular contexts and persons, and deal only with certain formal, semantic, and epistemological characteristics of sentences.

The replaceability of explanations means different things, depending on whether one envisages replacing a system of explanations merely in its logical derivation role, or replacing it in all its pragmatic uses. One of the conditions necessary for an eliminative-reduction is that the reducing theory logically imply all of the observation statements logically implied by the reduced theory. Furthermore, the reducing theory T must be more general, precise, simple, or systematized than T'. In that case we can reject the theory T' and withdraw commitment from its hypothetical entities; T' is no longer needed in order to give a logical derivation of the observation statements, and it can be replaced in that role by T.[2]

However, some authors make the far stronger claim that the reducing theory can replace the reduced theory not only in the context of logical derivation of observation statements, but in pragmatic contexts of explanation as well. A speech act using terms of the reduced theory can be replaced by one using only terms of the reducing theory, and still succeed in rendering the puzzling fact clear or intelligible. For example, Rorty says:

> If the Identity theorist is taken to be predicting that some day 'sensation,' 'pain', 'mental image', and the like will drop out of our vocabulary, he is almost certainly wrong. But if he is saying that, at no greater cost than an inconvenient linguistic reform, we *could* drop such terms, he is entirely justified. And I take this latter claim to be all that traditional materialism has ever desired.[3]

The crucial claim is that psychological terms could, at no greater cost than linguistic inconvenience, drop out of our vocabulary and leave our ability to describe and explain undiminished. In a later article, Rorty says:

> If we knew enough physiology, we could forget about commonsense ways of accounting for the behaviour of our friends and relations . . . common sense psychological theory is a dispensable crutch . . . if we

[2] Kemeny and Oppenheim define the elimination of the reduced theory in terms of its superfluity in contexts of logical derivation of observation statements. See their article 'On reduction'.

[3] Rorty, 'Mind-body identity, privacy and categories', pp. 198–9.

understand materialism as a possible *replacement* for common sense, then it is obviously true.[4]

Feyerabend agrees. He claims that 'there are enough independent predictions available, many more predictions in fact than the mentalist could ever provide' for physiology to provide its own explanatory system for behaviour. This system would be independent of psychological terms in the sense that its terms would not be related to those of psychological theories. For the physiologist who wishes to avoid the conceptual difficulties of the mind-body problem, 'the proper procedure for him to adopt is to develop his theory without any recourse to existent terminology'.[5] Feyerabend argues that a 'materialist language' could replace psychological terms and explain and describe the world better than our ordinary psychological language.[6]

In effect Rorty and Feyerabend claim that once reduction is achieved, whenever some puzzling fact is rendered clear or intelligible to a particular person in a particular context by an explanation using psychological terms, that explanation can be replaced by one similarly successful using only physiological terms. This replacement claim is far stronger than the claim that the reducing theory will logically imply every observation statement implied by the reduced theory.

The Replacement assumption makes a similarly strong claim about the pragmatic replaceability of particular explanations. It reads:

> If a theory T' reduces to a theory T, then any explanation using a term from T' can be replaced by some explanation using a term or terms from T.

While claims about derivation concern certain logical relations between sentences, this Replacement assumption—and the claims of Rorty and Feyerabend—concern the *uses* of those sentences in particular contexts of explanation. Where now we use psychological terms in our explanations, we could use only physiological terms.

[4] Richard Rorty, 'Functionalism, machines, and incorrigibility', *Journal of Philosophy,* LXIX, 8 (20 Apr. 1972), 219.

[5] Feyerabend, 'Comment: "Mental events and the brain"', in *The Mind-Brain Identity Theory,* ed. Borst, pp. 140–1.

[6] Feyerabend, 'Materialism and the mind-body problem', pp. 147–9.

I shall argue that this strong form of the Replaceability thesis is incorrect. In fact we have already seen reason to doubt even the weaker form of replacement: that physiological theory will *logically imply* all the observation statements currently derived from psychological theory. In order to deal with the possibility of loose localization of function one may require connecting principles which are disjunctive, non-specific, or incomplete. In that case, a logical derivation of the psychological theory from the physiological theory is precluded. We could not replace psychological terms with physiological ones in ordinary contexts of description or explanation, for we would not have even a physiological specification of the referents of the psychological terms. Yet if we accept certain inductive risks (see pp. 96–7), we could show that the nervous system is functionally isomorphic to the psychological model, and thereby achieve a model-reduction. The Replacement assumption need not be satisfied in all types of reduction.

However, let us suppose that functions are strictly localized, that we find general identities linking psychological and physiological terms, and that we can logically derive psychological theories from physiology. Would the stronger form of the Replaceability thesis then hold up?

A physiological explanation of stimulus-response relationships would require a physiological specification of what constitutes a given stimulus or a given response. Presumably one could give a physiological explanation of a particular behavior by tracing the sequence of neural events from reception of the relevant stimuli to generation of the response. One would need to know the determining events for the firing of each neuron in the relevant pathways, which would entail specifying the synaptic events of each of them. But assuming that the output of the system is a function of its input and its internal states, and that we can give a physiological specification of stimuli and responses, nothing in principle bars a purely physiological explanation of stimulus-response regularities.

I shall argue that even if we had such physiological knowledge and could derive our psychological theory from physiology, nevertheless we could not replace our psychological explanations in ordinary contexts of description or explanation with physiological explanations. Barriers other than mere linguistic inconvenience would prevent the replacement of explanations.

B. *Ineliminability of Psychological Explanations*

An example may help to reveal the barriers preventing replaceability of explanation. Studies of self-stimulation in rats have gone a long way towards localizing the neural structures mediating reward. Indeed, some investigators have claimed that the noradrenergic (NA) pathways are precisely the structures involved.[7] Although the evidence for this claim is not conclusive, let us suppose that reward is identical with the passage of excitation through the NA pathways of the medial forebrain bundle (MFB). Consider how this would affect the replaceability of explanations. Why do rats self-stimulate? One answer is that they self-stimulate because the electrode activates their mechanisms of natural reward, and so the probability of the rat pressing the bar increases, just as it would for any instrumental response. Another explanation of this behaviour is that the rat self-stimulates because the current releases noradrenaline in its MFB. In what sense could the latter explanation 'replace' the former?

The physiological explanation prompts one to ask: what does the release of noradrenaline in the MFB have to do with the bar-pressing behaviour? That is, even though that physical process is by hypothesis identical with reward, the purely physiological description does not make clear the connection between the physical process and the behaviour in question. It provides an explanation of self-stimulation only if one knows that the NA pathways of the MFB are the reward pathways, and that stimulation of those pathways therefore increases the probability of an immediately-preceding response. The psychological explanation, on the other hand, does not leave this gap in the explanation. 'Reward' is a term in part defined by its relations to behaviour change—an increase in the probability of an immediately-preceding response—and so citing 'reward' explains the behaviour.

[7] L. Stein and C. D. Wise, 'Amphetamine and noradrenergic reward pathways', in *Frontiers in Catecholamine Research*, eds. E. Usdin and S. H. Snyder (Oxford: Pergamon, 1973), pp. 963–8.

This is a general point. Part of the meaning of any psychological term is that the psychological state has certain relations to stimulus variables, to other internal variables, and to response variables; so that in some situations, change in that psychological state entails certain changes in behaviour. Even if we find some neurophysiological term which picks out just the values of a given psychological variable, the neurophysiological term does not carry as part of its meaning those relations to other internal variables and to stimulus-response relationships. Although in our hypothetical example, reward is identical to activity in the NA pathways of the MFB, the claim that there is increased activity in the NA pathways of the MFB does not entail the claim that the probability of an immediately-preceding response will increase; while at least in some situations, increase in reward entails a certain change in behaviour. Neural terms do not in general characterize states and processes in terms of their functional role in behaviour. They are defined in a different context, related to different observations (those of physiological states of the nervous system), and they carry different truth conditions and implications.

This simple point suggests why the thesis of replaceability of explanations is incorrect, even in cases where we find precise general identities between psychological terms and neural ones. Although one can explain why the rat self-stimulates by saying that it activates its reward mechanism, one could not in general successfully replace this explanation with a physiological characterization of the reward state. One needs to specify the relations between the stimulus conditions, internal states, and behaviour; and to do that one needs functional terms.

At this point an imaginary interlocuter might raise the following powerful objection: 'It is obvious that replacement of a psychological term by a coextensive physiological term will not suffice to explain the behaviour which is explained by use of the psychological term. Since the terms have different truth conditions and implications, they cannot fulfil the same jobs in contexts of explanation. But why could we not use a *collection* of physiological terms to explain the relation between the given internal state and the behaviour? The

Replacement assumption does not entail that a given psychological term is replaced in every context by a particular physiological term; it merely claims that a collection of physiological terms as a whole can do just as well in explanations as the collection of psychological terms. For example, we could presumably use other physiological terms to explain the relation between current passing through the NA fibres of the MFB and the self-stimulation behaviour of the rat. By hypothesis we can derive from physiological theory the same stimulus-response relations as derived from psychological theory. If this is so, however, we can simply trace the physiological effects of excitation in the NA fibres of the MFB from circuit to circuit until we explain its effect on motor behaviour. In that way a purely physiological explanation replaces the psychological one.'

The power of this objection is that it abandons the attempt to make a term-by-term replacement and allows one to use a collection of physiological terms to explain the relation of the given internal state to behaviour. However, I believe that it does not establish the possibility of replacement and succeeds only in underlining the distinction between the pragmatic aspects of explanation and logical-derivation aspects. Consider what is required in order to replace a particular psychological explanation in a particular context with a physiological explanation. The questioner is aware of certain facts, has certain assumptions, and wants some puzzling fact explained. In psychological explanations the questioner may be aware of certain stimulus conditions of the creature (and perhaps some of its internal states) and is puzzled over some aspect of the creature's behaviour. We explain the behaviour by citing some functionally-defined state connected with those stimulus conditions (and other internal variables) and with the behaviour, thereby clearing up the confusion.

For a physiological explanation to replace the psychological one, an explanation using only physiological terms must answer the same question, relying only on the same data. Otherwise the physiological explanation will not succeed in making clear or understandable the same puzzling phenomenon explained psychologically. If the physiological explanation relies on evidence which is not required by the psychological one and which is not in the possession of the questioner, then it will not succeed in replacing the psychological explanation.

This is why the use of a collection of physiological terms fails as a replacement for the psychological explanation. The questioner may be in possession of evidence which allows the justifiable ascription of a given functional state, but this same evidence will not be sufficient to ascribe the required collection of physiological terms. On the basis of the experimental set-up and the stimulus characteristics, one may be able to ascribe 'increased reward' to the rat, but that same evidence will not be sufficient to ascribe to it all the physiological terms characterizing the sequence of effects of excitation in the NA pathways of the MFB. To trace those physiological effects, one would need terms characterizing activity in many different circuits and structures (terminating in the motor circuits), and the evidence needed to ascribe those terms may include assays, micro-electrode readings, pharmacological data, and so on. In the context of the psychological explanation, one may not have any of that evidence. As a result, the psychological explanation may succeed in rendering some puzzling behaviour intelligible where its physiological replacement would fail.

It might be objected here that in some cases the observations sufficient to ascribe a functional state are also sufficient to ascribe a physiological state, since the two terms may refer to the very same state. 'Reward' by hypothesis refers to the very same state as 'excitation in the NA pathways of the MFB', and if we have grounds for ascribing one, do we not also have grounds for ascribing the other? This objection fails, however, for if knowledge of the identities linking physiological terms to psychological ones is necessary in order to ascribe the physiological terms, then the psychological terms play an ineliminable role in the physiological explanations. Knowledge of the functional role subserved by the given physiological state is still required, and the psychological terms therefore cannot be eliminated.

An example in another discipline—the identification of nucleic acids as the primary components of genes—may clarify this point. Gregor Mendel studied the patterns of inheritance of various traits in the garden pea, and proposed that these patterns could be explained by pairs of 'unit factors' which were sorted and combined in certain ways during reproduction. These unit factors or *genes* were at first merely functionally defined: pairs of them control the inheritance of a Mendelian trait, and their sorting during reproduction follows certain laws.

Only in the twentieth century were genes identified as portions of chromosomes, made up largely of DNA. A hypothetical entity which had been functionally defined was given a structural characterization, and with techniques of chromosome mapping it became possible to identify the particular chromosome sequence which is a gene for a particular trait.

The important point is that the term 'gene' gives a functional characterization of portions of chromosomes in terms of their role in inheritance of traits, while the biochemical terms are not linked in that way to that context. Even though genes are portions of chromosomes, the term 'gene' and the biochemical terms are defined by relations to different sorts of observations, and are useful in answering different sorts of questions. One may sometimes explain the pattern of occurrence of a particular trait by referring to the dominance or recessiveness of the gene or genes controlling it, and to its constellations of occurrence in parent generations. To replace the genetic terms in such an explanation with biochemical specifications of the gene would however puzzle one's listener, unless he knew that the particular chromosome portions picked out were the *genes* for the trait in question. The fact that genes are portions of chromosomes does not entail that all explanations which cite genes can be replaced with explanations which cite nucleotide sequences. Because a functional characterization of a state is different from a biochemical description, the former may clarify matters while the latter fails to do so.

Why is replacement precluded? Pragmatically, explanation consists of rendering some puzzling phenomenon clear or intelligible for a questioner on the basis of evidence he or she can accept. This requires terms which can be ascribed on the basis of the given evidence and which have implications concerning the phenomenon in question.[8] If two classes of terms have different observational criteria of ascription or different implications, use of one kind of term to answer a question in a particular context could not in general be replaced by use of the other kind of term. But functional and structural terms are related to different observations and have different implications. Hence the replacement of the former by the latter in contexts of explanation is precluded.

This argument renders the replaceability thesis dubious for *any* case of reduction involving theoretical terms from two different experimental disciplines. Theoretic terms gain a use through their inferential connections to experimental procedures and observations, and they are used to explain other observations; theoretic terms from different disciplines will be inferentially linked to different sets of observations and procedures. But as long as the uses of the two sets of terms are in these ways distinct, one of the sets cannot take over the explanatory and descriptive roles of the other.

For example, the term 'temperature' is given its meaning in thermodynamic contexts, where it is related to measurements, thermal perception, thermometer readings, and so on. Even though the temperature of a gas is the mean kinetic energy of its molecules, the concept 'mean kinetic energy of molecules' is associated with observations in a different experimental context. As long as the truth conditions of the terms inferentially relate them to such different sets of observations, and establish their applicability in such different ways, the term 'mean kinetic energy of molecules' could not fulfil the descriptive and explanatory uses of 'temperature'. To take over those uses, the theoretical term 'molecule' would have to be inferentially related, in the same way as 'temperature', to the observations and procedures of thermodynamics.

This is not to deny that 'temperature' and 'mean kinetic energy of molecules' come to be related, to a certain extent, to some of the same observations. Although reduction is likely to change the meanings of some theoretical terms, the thesis

[8] Accounts of the pragmatics of explanation are much more complicated than this simplified formula might suggest. See Sylvain Bromberger, 'An approach to explanation', in *Analytical Philosophy*, ed. J. R. Butler, 2nd ser. (Oxford: Basil Blackwell, 1965).

of complete replaceability in pragmatic contexts must, I think, be abandoned. It is a tautology to say that classes of terms from distinct disciplines will be related to distinct contexts of explanation and will have their own roles to play within inquiry; that they will be useful in answering different questions. Certainly there are instances of one theory reducing tc another, without the disciplines producing those theories thereby collapsing into one. I suggest that in any such case, the thesis of replaceability fails.

The crucial point is that the conditions defining explanatory reduction concern the logical relations between sentences, and do not depend upon the explanatory or descriptive uses of those sentences. That is, even though sentences *P* may be derivable from sentences *N* conjoined with connecting principles *C*, this does not show that the explanations using sentences *P* can be replaced by explanations using sentences *N*. The distinction here is as simple as the distinction between a sentence and its use; for while reduction is a matter of logical relations between sentences, replacement concerns the explanatory uses of those sentences. Reduction has no direct implications concerning the explanatory uses of the sentences.

Philosophers have sometimes thought of the relations between different levels of explanation as antagonistic, so that success of the reductionist programme would mean that one kind of explanation invalidates or eliminates the other. In fact, the inter-derivation of sentences does not entail that the two systems of explanation collapse into one. We should replace the image of antagonism between different levels of explanation in inter-theoretic reduction with the image of mutal coexistence between parties with slightly different concerns. The different realms of discourse as represented by the two disciplines have their own puzzles, their own typical ways of answering those puzzles, and their own concepts useful in answering them; and these features of the disciplines can remain independent and coexistent even though theories in the disciplines may be inter-derivable. The differences between the roles of terms in inquiry can survive a reduction of one theory to another.

2. *Incompatibility of Theoretic Frameworks*

According to eliminative-reduction, one reason why the reducing theory replaces the reduced theory is that the two theories contradict each other in some experimental predictions, and furthermore the theoretic framework of the reducing theory contradicts some of the presuppositions of that of the reduced theory. In such a case there is a genuine incompatibility between the two systems of explanation. Will the reduction of psychological models to physiology show this feature?

The analysis strongly suggests that it will not. If we show that the nervous system is functionally isomorphic to a psychological model, then far from abandoning commitment to the existence of the hypothetical states and processes in the psychological model, we actually give them a physiological identification. Reduction justifies our entering existence claims for comparators, filters, expectancies, safety signals, and the like; for establishing functional isomorphism shows that there are parts and states of the nervous system corresponding to those parts and states in our model. It is only if reduction fails that we withdraw existence claims from our hypothetical psychological states, for such a failure shows that there are no neural states fulfilling the given functional role.

If the psychological model and the nervous system are functionally isomorphic then they will both apply to the same stimulus-response relationships. Any inconsistency between their predictions is reason to adjust the psychological model to achieve a match, and not to abandon it. The goal is a unified psychological-physiological account, and hence the claim that the reduced and reducing theories have contradictory experimental predictions is incorrect.

We also have reason to doubt that provision of physiological explanations contradicts the presuppositions of our psychological explanations. I pointed out in Chapter II (p. 55) that one can give psychological models for simple goal-directed behaviour in animals based on the idea of a negative feedback device, and that parts of the nervous system are instances of such devices. This model suggests a resolution to a supposed incompatibility of physiological and psychological explanations, by demonstrating how (admittedly very simple) forms of goal-directed behaviour can be manifested by a system all of whose parts obey the laws of biophysics. It thereby shows that there is not a direct incompatibility between saying that causal laws describe all of the activity of the system and saying that the system shows goal-directed behaviour. There is not a conceptual barrier, impervious to all growth in knowledge, preventing us from relating goal-directed behaviour to neural functioning. Perhaps the problem is one of devising psychological models of sufficient complexity.

Furthermore, such a model licenses continued use of explanations citing goals; the psychological states cited in such explanations are given a physiological identification, and their relations to particular neural structures can themselves be explained. We explain how a neural state can be the given psychological state by showing how it can play the appropriate functional role in the system. Paradoxically, therefore, the provision of a neural mechanism for goal-directed behaviour confirms the reality of goal-directedness and the propriety of explanations in terms of goals. Discovery of an adequate psychological model for the capacity shows how a system of neurons could fulfil the functions required, and hence dissolves the supposed incompatibility between psychological and physiological explanations.

3. *Meaning Change*

A subtle defence of the replaceability thesis has yet to be considered. One can claim that reduction results in a change in meaning of terms in the reduced science, so that its explanations are in effect converted into explanations of the reducing science. The assumption that terms of the two disciplines involved in inter-theoretic reduction are related to different observations and have different implications was important to the conclusion of the argument in the first section of this chapter. However, do not those criteria and truth conditions change as a result of reduction? Meaning change consequent upon reduction may reduce the semantic differences between structural and functional terms.

According to Feyerabend's analysis of reduction (see above, pp. 26–8), sometimes the vocabulary and (modified) laws of the reduced theory are retained, as in the retention of a modified Newtonian physics even after its reduction to Einsteinian laws. However, when it is retained, the reduced theory undergoes certain crucial alterations: its theoretical concepts change in meaning, so as to be accommodated to the different theoretical framework of the reducing theory. Although Newtonian physics seems not to have been eliminated or replaced by the Einsteinian variety—in the sense that Newtonian theoretical terms are retained—this appearance is deceptive, for the retained portions of the theory have changed in meaning. The concepts of 'mass', and 'space', and 'inertia' have new, relativistic meanings, and are now part of relativity theory.

In the section which follows I shall explore some of the meaning changes which can plausibly be expected from reduction in psychology, to see if this general thesis holds up.

A. *Change in Truth Conditions*

One kind of meaning change in reduction is unproblematic: given a general identity between a psychological and physiological term, the two terms can exchange some of their truth conditions or criteria of application. (By 'criteria of application' I mean the kinds of evidence relevant to, yet perhaps neither conclusively nor deductively establishing the justified ascription or denial of a term.)

For example, if depression really is a malfunction in reinforcement mechanisms—specifically, a depletion of noradrenaline in limbic reward pathways—then the physiological and the psychological terms could contribute to each other's truth conditions. Noradrenaline depletion in one's limbic reward pathways could come to function as evidence for the claim that one is depressed; 'depression' could come to absorb as part of its truth conditions 'being NA depleted in limbic pathway A6'. In this way, a theoretical term in a psychological model gains a physical connotation, so that from the physical state one could infer the depression.

We can point to many examples of physical states found to be constantly conjoined to psychological states and coming to function as evidence for their ascription. The use of rapid-eye-movement sleep as a criterion for dreaming in humans is one example; pupil dilation as a measure of degree of neural activation (hence arousal) is another. These inferences are no more problematic than any inductive inference. As scientific knowledge advances, previously unrelated phenomena are found to be related; and given the success of the general theory ascribing a correspondence between a psychological and a physiological state, inferences from one to the other would be justified.

Perhaps a more interesting possibility is that of inferences from psychological states to neural conditions; that is, given the identification of a psychological variable with a class of neural states, evidence for ascription of that psychological state can also come to justify the claim that one is in a particular physical state. If in some future psychopathology depression is successfully identified as NA depletion in limbic pathways A6, then whenever one has grounds for thinking someone is in a state of reactive depression, one also has reasonable grounds for believing that that person's nervous system is in a particular condition. If I come to think that I myself am depressed, then I also have grounds for thinking my limbic pathway A6 is NA depleted.

To both of these kinds of inference, the following objection can be raised: although one can perhaps come to make inferences from a neural condition to a psychological state, and from a psychological state to a neural condition, in both cases those inferences employ the general theory linking the two states as a tacit premiss, and so the two terms have not exchanged any of their truth conditions, as is asserted to happen in instances of meaning change. Instead of saying that depression is sufficient grounds for the inference that one's limbic pathways are in a certain condition, we should describe the situation by saying that we now know more about depression, and our new knowledge functions as a premiss in all such inferences.

My response to this objection is simply that no one has yet managed to distinguish precisely between change in the meaning of a term and increased knowledge about the referent of that term. One can support the claim that the terms 'depression' and 'NA depletion in pathways A6' retain their distinct meanings only by claiming that in all the inferences from one to the other, the theory functions as a tacit premiss; and here one must admit that the degree to which the theory can, as a premiss, be left merely implicit, is quite large. The background knowledge is rather far in the background, in some cases. If the two terms come to be used in such a way that the general theory is never cited as a premiss in inferences from one to the other, then there is little point in insisting that nevertheless the theory must be employed as a tacit premiss. It is more parsimonious to assume that the terms have exchanged some of their truth conditions or criteria, so that inferences between them are direct.

B. *Role of Theoretical Terms*

'Hunger' is a typical internal variable in a psychological model, defined in terms of certain relations to antecedent variables such as hours of food deprivation or metabolic activity, and to consequent variables such as food-seeking behaviour, reinforcement for instrumental responses, or general arousal. These relations define what it means for the rat to be hungry, so that in certain situations, particular behaviours provide evidence for the ascription of 'hunger'. Just like a theoretical term in other sciences, evidence for the ascription of the term is indirect, and rests on presumed relations to various kinds of observations.

Suppose that in a developed model of feeding behaviour we identify hunger with stimulation of glucoreceptors in the hypothalamus, where the glucoreceptors are identified by certain strucural chacteristics. (This over-simplified model of feeding was abandoned fifteen years ago, but I cite it just to illustrate the logical point.) We then need only to insert microelectrodes into the appropriate neurons, and our recordings will give us direct evidence as to whether or not the rat is hungry. We shall no longer need to rely upon indirect inferences from behaviour in particular situations. Once the term 'hunger' is pinned down to the stimulation of particular structures, its epistemological status seems to change radically.

In effect, what seems to happen is that a theoretical term

becomes 'less theoretical', in the sense that evidence for its ascription was formerly indirect, whereas now some of it is relatively direct. The theoretical term becomes linked to new kinds of observations—recordings from the glucoreceptors—which seems to involve a change in meaning of the theoretical term. Since we would have relatively direct evidence for the hunger state of the rat, we could investigate other properties or characteristics of the state, independently of the behavioural evidence.

Certainly these changes in the use of 'hunger' could result from a successful mapping of a model on to the nervous system. However, they are not peculiar to eliminative-reduction, and are found in derivation-reduction as well. Given the establishment of the connecting principles, terms in both the reduced and the reducing science are related to new sets of observations, and these relations can provide new grounds for ascription of the terms. Furthermore, theoretical terms can thereafter be related relatively directly to observational evidence. For example, with the reduction of thermodynamics to statistical mechanics, according to Nagel the molecular state of a substance was linked to new kinds of experimental evidence: its temperature and pressure. The theoretical term 'molecule' was linked to new kinds of observations. Given the successful reduction, if I perceive that my electric kettle is warming up, I can justifiably infer that the kinetic energy of the molecules inside it is, on average, increasing. The term 'molecule' seems to become less theoretical in the sense that it is tied relatively directly to new observations.

A final example perhaps provides a closer analogy to reduction in psychology. While Mendel's 'unit factor' was defined only functionally, in some cases the particular nucleotide sequences which are particular genes have been identified. With this development the term 'gene', formerly linked only with patterns of trait inheritance, has now in many cases been given a biochemical specification. In some cases some of the protein structure of simple organisms has been read off the organism's chromosomes.[9] 'Gene' discourse, while formerly relatively theoretical, is now much more directly tied to

[9] See F. Sanger *et al.*, 'Nucleotide sequence of bacteriophage ϕX174 DNA', *Nature*, 265 (24 Feb. 1977), 687–95.

observations, and it is somewhat inappropriate to call a gene a hypothetical entity at all. The ability to read off some of the genetic characteristics of an individual is perhaps analogous to determining, in a future physiological psychology, the value of an internal psychological variable with a micro-electrode or some other physiological procedure.

C. *Analogies*

To this general line of argument, one can raise the following objection. When one says 'The molecules in the kettle are moving faster' instead of 'The water is getting hotter', one is making a joke by using theoretical discourse in an odd way. The only grounds for saying 'The molecules are moving faster' is that one feels the water heating up, and all one means to say with this theoretical expression is that the water is getting hotter. This meaning is conveyed to one's listener only if he or she knows something about molecules and heat. In that way, the sentence 'The molecules in the kettle are moving faster' is merely grafted on to 'The water is getting hotter'. It is useful in this context only if the hearer knows the relation between kinetic energy of molecules and heat. He must infer that by 'The molecules in the kettle are moving faster' one means 'The water is getting hotter', and must base this inference on his knowledge of the relation of the two phenomena. But if an inference is required, then the two terms have not traded the relevant truth conditions.

There *is* something odd about saying 'Look at the electrostatic discharge!' instead of 'Look at the lightning!'; or 'Please pass the sodium chloride' instead of 'Please pass the salt'; and this oddity is something like the oddity of saying 'My glucoreceptors are being stimulated' instead of 'I am hungry', or 'My limbic NA pathways are depleted' instead of 'I am depressed'. In all of these cases we seem to tie a theoretical term very tightly to observations. We apply the term solely on grounds of the observation and the general theory, and we convey our meaning only if our listener knows that theory. 'My glucoreceptors are being stimulated', conveys to one's listener the fact that one is hungry only if the listener knows that the hunger state is just a state of stimulation of the glucoreceptors. One's evidence for the assertion that those neurons are being stimulated is just that one is hungry.

However, it is false to say that these uses of the theoretical discourse are *merely* metaphorical, and that all that is meant by saying 'The molecules in the kettle are moving faster' or 'My glucoreceptors are being stimulated' is that the water is getting hotter, or that I am hungry. The theoretical terms are not in these cases merely grafted on to the observation terms, but have a meaning of their own.

In particular, if the water is heating up, and belief in the general theory is well founded, then when one says 'The molecules in the kettle are moving faster', one is actually making a well-founded assertion about the molecules in the kettle; it may be literally true. Similarly, we can deny that the physiological vocabulary used in these contexts gains its meaning entirely by analogy, because when one uses the physiological term instead of the psychological one, one is making an assertion which may be literally true. When I say 'My limbic pathway A6 is NA depleted', not only does the hearer come to know that I am depressed—supposing he knows the general identity claim—but also I may be making a true assertion about the state of my nervous system.

In this way grounds for ascribing psychological states to oneself can also function as grounds for ascribing certain states to one's nervous system, and one can use physiological sentences to make avowals about the psychological state of oneself or of others, given that one's hearer also knows the relevant identities.

We can clarify this by considering some descriptions of bodily states, used to communicate psychological states, which are merely analogies. For example, although 'He is hot-blooded' was once taken to describe a physiological condition (the nature of his blood) which explained some of his behaviour, with the abandonment of that particular model of human character the sentence became merely a metaphor. In saying he is hot-blooded, we do not mean to assert that his blood is of a certain temperature. Similarly, it was once thought that romantic distress affected one's heart in certain ways, and that because the heart was the organ of the emotions, this change explained one's behaviour; although now when we say

'My heart is broken' we are certainly not describing a vascular accident. In effect, the sentence once played a role in an unsophisticated physiological psychology, but is now used merely as a picturesque metaphor.

Interestingly enough, we do have certain terms describing emotional states whose status is ambiguous between metaphor and genuine physiological characterization. For example, if I say 'I need a drink to calm my nerves', it is unclear whether the reference to 'nerves' is mere metaphor. The evidence that minor tranquillizers depress the activity of the septal-hippocampal circuits thought to be involved in behavioural inhibition, fear, and anxiety,[10] makes one think that perhaps the calming effect on the nerves can be taken literally. Similarly, if I say 'He had a nervous breakdown', the quasi-mechanical image this suggests may not be entirely metaphorical, as presumably, at least in some cases, there is some neural malfunction to which 'nervous breakdown' refers. Other examples are 'stimulating' lectures (stimulating certain neurons?), 'energetic' executives (high amounts of potential energy in their motivational pathways?), 'irritable' employers, 'sensitive' sophomores, and 'impulsive' adolescents.

D. *Structural Terms and Functional Terms*

I have said that we can use physiological characterizations to communicate our psychological state (as in 'My glucoreceptors are being stimulated' instead of 'I am hungry', or 'I have alpha rhythms in my neo-cortex' instead of 'I am drowsy'); but that in all such cases, communication is effective only in so far as the hearer knows the relation between the physiological and psychological conditions. This raises the problem, noted earlier, of distinguishing inferences based on tacit premisses from changes in the meaning of terms. If the physiological characterizations can sometimes play the role of the psychological terms, could they absorb all of the truth conditions and criteria of the psychological terms, and simply replace them altogether? In that case, the hearer would not need to infer one's meaning, since 'My glucoreceptors are being stimulated' would imply what is now implied by 'I am hungry'. Perhaps

[10] Gray, *The Psychology of Fear and Stress,* pp. 152–6, 207.

as knowledge of the general identities spreads, such usage will become more common, and physiological terms will gradually take over all the descriptive and explanatory roles of our current psychological vocabulary.

In this way meaning change reopens the possibility of eliminative-reduction. I said above (p. 164) that for physiological explanations to replace psychological explanations, the former must answer the same questions, relying on the same data and assumptions as the latter. This requires in turn that the collection of physiological terms absorb the entire framework of functional characterizations associated with psychological terms, so that the physiological terms could be ascribed on the same grounds and would be related to behaviour in the same way as the psychological terms. We would describe the nervous system with just one kind of term, which would both identify structures and characterize their functional role. Terms such as 'motor cortex', 'reticular activating system', and 'reflex arc' are perhaps examples of terms which can do both jobs. If the terms can exchange some truth conditions and criteria, perhaps the physiological terms could absorb all of the truth conditions and criteria of the psychological terms. Could we not then eliminate our psychological explanations and replace them with physiological ones?

The first point that should be made about this prospective elimination of psychological terms is that they are eliminated only in name. That is, if physiological terms must absorb the entire set of functional characterizations given by the latter as part of their truth conditions and criteria, then functional concepts are not eliminated; they are simply given a new label. The new label is 'physiological' instead of 'psychological'. Unlike genuine cases of eliminative-reduction, here the reducing theory would not provide a radically new, incommensurable theoretical framework for the observations of the reduced theory; it would simply relabel the old one.

Secondly, there are serious drawbacks to uniting structural and functional characterizations of a system into one set of terms, and it seems quite unlikely that such a unification would ever serve to advance theory. The semantic differences between the concepts seem to preclude using only one term to characterize structures and functions. There is a difference between asking what a part in the black box does, and how it does it; between identifying a part in terms of its input-output relations, and identifying its physical make-up and organization; between describing the relations of the activity of a part to other parts and to the system as a whole, and describing its constitution and internal workings. A unification of vocabulary would need to reconcile these semantic differences.

Furthermore, if we could generate unified structural-functional terms, they would be heir to some difficult ambiguities and weaknesses. Different structures can fulfil a given function, yet our proposed unified terms could not easily describe this possibility. The term 'motor cortex', for example, is ambiguous. Does it refer to a particular area of cortex (the precentral gyrus) which happens to have motor functions? Or does it pick out those areas of cortex subserving motor functions (which, in the normal individual, happen to be the precentral gyrus)? This ambiguity becomes a problem if we ever have to deal with an abnormal case, in which the precentral gyrus does not subserve motor functions. Should we say then that the motor cortex is some *other* area of cortex? Or should we say that the motor cortex in this case happens not to subserve motor functions? Or perhaps the individual has no motor cortex! In order to deal with the possibilities of loose localization of function, individual differences, or change in function after lesions or disease, we need to characterize the function of a given part with terms distinct from those used to describe its make-up and organization.

As long as there is a semantic difference between structure concepts and function concepts, the meaning change in reduction cannot proceed so far as to eliminate the need for psychological terms.

This argument may be clarified if we contrast it with a weaker position. It may be thought that psychological terms (in the sense of terms used by experimental psychologists) are ineliminable because psychology and physiology are distinct disciplines, and therefore terms used by psychologists are related to different observations, experimental techniques, and behavioural implications from those of physiologists.

I think that this position is incorrect, and that parts of psychology and physiology will in some sense merge. For example, experiments aimed at localizing functions utilize experimental designs and techniques from both psychology and physiology. Experimenters in physiological psychology use both behavioural and physiological terms, and such terms are related to both behavioural and physiological observations. The terms used by psychologists are not distinct sets, but increasingly overlap, with psychologists using 'physiological' terms, and physiologists using what we now call 'psychological' terms. The disciplines of psychology and physiology are joined by a grey zone of physiological/psychological experiments, combining techniques, vocabularies, and observations of both disciplines.

If the argument against meaning change were based on a presumed dichotomy between psychological and physiological vocabularies, techniques, and observations, then it would fail. However, my claim is that even in experiments aimed at localizing functions, and employing both psychological and physiological vocabularies and observations, one can pick out two classes of term. These two classes play different roles in inquiry, are used to answer different kinds of question, and have different criteria and truth conditions. The different roles of the terms can be marked by calling one sort 'functional' and the other 'structural'. The important distinction is not between the current vocabularies employed by psychologists and physiologists, but is rather the distinction between the semantic features of function concepts and those of structure concepts. The use of functional terms in explanations cannot be replaced by structural terms. This is why what we now call psychological terms are ineliminable and play an irreplaceable role in inquiry. As long as there is a difference between asking what a part does and asking how it does it, we shall find psychological (functional) terms employed.

In conclusion, even though a psychological model can be reduced to physiology, such a reduction does not entail an eliminability or replaceability of the psychological vocabulary. Barriers other than linguistic inconvenience would prevent this replacement. The psychological and physiological terms have different sorts of uses, answer different sorts of questions, and characterize the system in different ways. This does not preclude an explanatory reduction; indeed, it opens the possibility of explaining why a particular psychological function is associated with a given neural structure, and it shows the basic compatibility of the two kinds of term. I accept the eliminative-reduction idea that the meanings of terms in the two disciplines can change; the psychological term can come to absorb as part of its truth conditions physiological indicators, and from the evidence for ascribing a particular psychological state one may be able to infer the presence of a particular neural condition. One may even be able to use physiological sentences to describe some of the conditions formerly described only with psychological terms. But this description relies upon the general theory and upon knowledge of the functional role of the physiological state. The change in meaning cannot proceed so far as to allow a replacement of psychological terms with physiological ones. One class of term describes *what* a part does by relating it to other internal states and to the input-output relations of the entire system; the other class of term describes *how* a part works by characterizing its physical make-up and structure. For this reason the use of a functional term in a particular context to clarify or explain a particular puzzle cannot be successfully replaced by a structural term. Although reduction can occur, it will not establish the eliminability of the psychological vocabulary.

I suggest that these features of reduction in psychology preserve what we could call the 'autonomy of the mental': they show why psychological terms must be retained in explanations, and why they cannot be replaced by physiological terms. Yet at the same time, this analysis shows that the psychological theories can be reduced to physiological theories, that the psychological and physiological terms are not incompatible, and that the relations of a psychological capacity to physiological structures can be explained. Psychological explanations are not autonomous in Putnam's sense—irreducible to physiological explanations—yet they are ineliminable, and retain a distinct role in inquiry.

Conclusion

It may be useful to review the course of the argument, and thereby summarize my analysis of the features of explanatory reduction in psychology.

Relating the terms in one discipline to those in the other is part of the problem of explanatory reduction, and the only kind of statement which currently relates psychological to physiological terms is a statement of localization of function. In order to understand the meaning of these statements one must understand something about model building in psychology. Such model building is intended to explain well-attested stimulus-response relationships; and to do this one proposes machinery inside the black box, which does the requisite jobs to generate a behavioural output from the stimulus input. One can specify those mechanisms in various ways; the important point is that mechanisms inside the black box are functionally characterized. Parts are specified in terms of their influence on other parts, and (thereby) on the input-output relations of the system as a whole. They are given 'jobs' in the sense that each part relates its inputs to its output; its job is defined in terms of its contribution to the stimulus-response relationships of the system as a whole.

Consideration of the way models are used in theoretical activity supports the idea that they functionally describe mechanisms in the creature which are responsible for its behaviour. The logic of discourse about internal variables in a model supports this point. Such discourse is theoretical, and takes its meaning from its relations to observable stimulus-response relationships. The necessity of deciding between rival models implies that each variable makes some detectable difference in behaviour. In some situations one must try to establish that the only difference betweeen the control and the experimental groups lies in variation in one internal variable; to do that one requires some well-attested connections among internal variables and between internal variables and stimulus-response relations. All of these points support the conclusion that terms for internal variables give a relational characterization of their values. These relational attributes provide the referential force of psychological terms and are true by definition of any psychological state. A necessary and sufficient condition for a state of the system to be a value of one such internal variable is that it bear the appropriate relations to other states of the system.

This analysis of what it means to say that a black box model gives a functional characterization of the mechanisms responsible for the input-output relations of the system is fundamental to the remainder of the argument. The idea that the job of a part in the hypothetical mechanism proposed by a black box model is a $function_c$ of that part—a set of consequences which the part has upon other parts and hence upon the activity of the whole—helps to clarify the conditions which must be met if a psychological function is to be localized in the nervous system. It suggests the conditions necessary and sufficient for reducing a black box model to physiology: that neural arrangements in the nervous system be found corresponding to every part of the hypothetical mechanism, and thus that every psychological function specified in the model be localized in the nervous system. The named structures and states of the nervous system must bear relations to each other which are isomorphic to the relations between the boxes and variables in the model, and this, I suggest, is the meaning of 'functional isomorphism'. For a psychological model to be reduced to physiology, the model must be functionally isomorphic to the nervous system. This claim is an implication of the idea that the connecting principles relating psychological terms to physiological ones will be statements of localization of function.

The analysis of functional characterizations of mechanisms inside the black box has additional consequences. It can easily be expanded to deal with some of the peculiar logical features of functional isomorphism: for example, the debate over strict versus loose localization of function; the possibility of disjunctive, non-specific, or incomplete connecting principles if functions are loosely localized; the imprecision of localized functions; and the way in which the connecting principles are interconnected, and succeed or fail as a whole. The use of statements of localization of function as connecting principles clarifies some of the peculiarities of the reduction, such as the possibility of opening up the basic parts in the psychological model, and the use of functional terms in physiology. Finally, the analysis has implications concerning the various types of reduction possible: the need for general identity statements in reduction, the goal of a unified model of the functioning of the nervous system, the validity of some functionalist anti-reductionist arguments, and the irreplaceability of psychological explanations.

A preliminary answer to the problem of reduction in psychology is that reduction succeeds if and only if one finds a psychological model functionally isomorphic to the nervous system. This condition is stated in terms of relating a *model* to the nervous system, however, leaving open the question of the equivalence or non-equivalence of model-reduction and derivation-reduction. The reduction of psychological models to physiology will prima facie show some of the features emphasized by the model-reduction account. One finds a substantive model in one discipline for phenomena described by another. Features of this model drop their status as mere analogies and come to represent real attributes. One uses identity claims to relate the two disciplines, and these connecting principles can themselves be explained. I argued that none of these features suffice to show that there is an instance of model-reduction which is not also an instance of derivation-reduction. However, other features of the case open this possibility: to show that the model is functionally isomorphic to the nervous system one does not require general identity claims as connecting principles. In cases in which functions are not strictly localized, the connecting principles may be disjunctive, non-specific or incomplete. One may successfully show that the nervous system is functionally isomorphic to the model, even though one cannot satisfy the conditions for the Identity version of derivation-reduction. Such a case would be an instance of model-reduction which is not also an instance of derivation-reduction using general identities as connecting principles.

General identity statements relating psychological to physiological terms may or may not be found; my main criticism of the Identity account of derivation-reduction is that such general identities are not required for reduction, and that that account therefore misrepresents some of the logical possibilities involved. A sufficient condition for a state to be a value of a psychological variable is that it bear the appropriate relations to the antecedent and consequent variables of that variable; different neural states could do this at different times. That the state satisfies a particular functional description is also a necessary condition for it to be a psychological state. Type-type identity laws therefore hold true only as long as the functional isomorphism of model and nervous system is preserved. Some versions of the Identity theory misrepresent the referential features of psychological terms; general identities will if anything be secondary to a demonstration of isomorphism of function between the model and the nervous system, for it is the latter that provides necessary and sufficient conditions for establishing the former.

The idea of giving a functional characterization of the mechanisms in the black box also casts light on eliminative-reduction. That psychological terms specify their referents with a cluster of behaviourally-defined relational attributes, while physiological terms usually do not, suffices to show that strong forms of the replaceability thesis are implausible. Barriers other than mere linguistic inconvenience prevent physiological terms from taking over all of the descriptive and explanatory roles of psychological terms. To acquire those explanatory uses, the physiological terms would need to absorb the framework of functional characterizations given by psychological terms. Although some change in meaning of physiological and of psychological terms is to be expected as a consequence of reduction, and a grey zone joins the two disciplines, the differences between function concepts and structure concepts will survive a reduction. Physiological terms can include function concepts when one investigates the inner workings of the basic parts in the black box model in order to see how *they* work, but relative to psychological terms and explanations of stimulus-response relationships, physiological terms convey *structural* characterizations of the system. As long as there is a distinction between asking *what* a part does and asking *how* it does it, it is impossible that one set of terms could take over the explanatory roles of the other.

I began the argument by sketching some deep-seated perplexities and worries concerning the prospect of reductionism in psychology, and it may be useful to re-examine them in the light of the completed analysis.

The example of reactive depression reveals some of these problems and perplexities. I showed how psychological and physiological explanations of depression employ different concepts, cite different observations and intervening processes, and have different implications. They seem to conceptualize depression in different ways and answer different questions. One explains a behavioural syndrome (with particular dispositions, inclinations, changes in abilities, and so on) by showing how certain learning phenomena influence motivational, cognitive, and emotional processes. The other explains a deficiency of transmitter substance in limbic circuits by showing how certain events cause a malfunction in particular pathways and structures.

It seems that psychological and physiological explanations derive from different conceptual frameworks, that they explain different things when they explain depression, and that therefore depression cannot be assimilated to both explanatory schemes. Depression is described correctly either by the psychological account, or by the physiological account, but not by both. The success of one rules out the other. We seem forced into a dilemma: either physiological reductionism is impossible, or, if it succeeds, it will invalidate and eliminate our psychological accounts, radically changing our conceptions of behaviour.

I argued that this dilemma is a false one. The differences between the explanations are reasons to deny that we could *replace* one sort of explanation in a given context with the other sort, for they cannot both be successful in explaining the same fact on the basis of the same observations. Yet that precludes them both from applying to a given behaviour only if we assume that whenever two theories provide explanations for a given domain, explanations provided by the reducing theory can *replace* those provided by the reducing theory. The analysis presented in this book provides a counter-example to this Replacement assumption.

In the last chapter I showed that even though psychological models can be reduced to physiology, nevertheless explanations employing psychological concepts cannot in general be successfully replaced by physiological explanations. Reduction can be achieved even though psychological and physiological explanations employ different concepts, are useful in different contexts, and proceed in different ways. We cannot in general replace functional terms with structural ones, for the terms have different observational grounds and different implications, and so are useful in different contexts for explaining different facts. The Replacement assumption must therefore be abandoned.

Once we realize that the explanations provided by reduced and reducing theories can be of different kinds, many of the problems associated with reductionism are resolved. For example, one of the supposed barriers to reducing psychological accounts of depression to physiological accounts is that the two explanatory systems employ different *concepts*: their terms are defined in different contexts and have different implications. The reduction of models of animal motivation and learning shows that this is not a good reason for denying the possibility of reduction. Functional concepts and structural concepts have different observational criteria, truth conditions, and implications, but a functional characterization of mechanisms in the black box can be related to underlying neural structures. Differences between the concepts do not preclude a reduction.

Psychological and physiological concepts can be related to one another in a unified model of the functioning of the nervous system. The various hypothetical states and processes invoked in psychological accounts are not abandoned in reduction; instead they are given a physiological identification. The presuppositions of the psychological terms are not contradicted by increasing physiological knowledge; instead that knowledge allows us to explain how it is possible that neural states can be psychological states.

Explanatory reduction is a matter of showing that some

sentences can be logically derived from others, and it has nothing directly to say about the explanatory uses of those sentences. Although one set of sentences can be logically derived from another set, this does not rule out the possibilities that the corresponding explanations employ different concepts, have different evidential grounds, answer different questions, and satisfy different criteria of adequacy. In effect we must distinguish claims about explanatory reduction from claims about the pragmatic aspects of explanations in the two disciplines. The Replacement assumption must be abandoned, for while explanatory reduction is defined in terms of the logical-derivation aspect of explanation (certain logical relations between sentences), replacement is a matter of the pragmatic aspects of explanation—the uses of those sentences in particular contexts.

I suggest that the unity of science is not a matter of homogeneity in the pragmatic aspects of explanation, but is rather a matter of interconnecting the sentences comprising theories of different disciplines. Different disciplines contribute different portions of this interconnected web of sentences. They may employ variously different experimental approaches and techniques to arrive at their theories, and they may use those sentences in specific contexts of explanation in variously different ways, for they are answering different questions and employing different terms. Various sorts of discourse may be employed in different disciplines, and the connecting principles between them may be of different varieties. What unifies all this disparate activity and different theoretical vocabularies is the goal of connecting the sentences comprising the theories into one coherent structure.

Reduction does not contradict the diversity of conceptual frameworks, experimental approaches, or types of explanation used in different fields; it is simply a matter of interconnecting sentences, allowing inferences from one set to another. Such logical derivation between theories in different disciplines must be sharply distinguished from other aspects of the disciplines—concepts and observations employed, experimental approaches, features of the pragmatic aspects of explanations, and so on—which can survive reductions. Interconnecting sentences is a relatively colourless affair, and carries few if any of the portentous consequences of the *bogeyman* of reductionism.

The analysis leaves a number of questions unanswered. One limitation is that it is confined to psychological models in a relatively circumscribed portion of the discipline: studies of motivation and learning in animals. One must wonder whether models in different portions of the discipline have the same general form as those considered, or whether other sorts of black box models must be included. One sort of theoretic vocabulary that has not been analysed in this book is that of information and information processing, and the relations of such terms to function discourse need clarification. Likewise, network models and procedural models have not been explicitly considered. An adequate account of reduction in psychology would need to consider the models in different portions of the discipline and the different uses of terms in those models.

A second question which could bear considerable elaboration is that of the relation of functional change to behaviour change. I considered very briefly some of the conditions required in order to make choices between models on the basis of experimental data, the role of controls in experiments, and the need for inferences linking stimuli or responses to internal variables. It was on the basis of these considerations that I concluded that psychological terms specify physical states in terms of the roles they play in the genesis of behaviour, and that such terms are defined by relational or functional attributes linking them to other internal variables and to stimulus and response variables. While I think this basic conclusion is true, it is perhaps unduly schematic. The complexities of claims about the meaning of theoretic terms are not adequately considered. The relations of psychological terms to observables and the use of controls display complexities and subtleties which the claim does not clarify. What is needed is a detailed study of the structure of arguments for or against particular models. Only a detailed study of the uses of psychological terms in those contexts could fully substantiate a functionalist account of their truth conditions and criteria, and such a study would no doubt make the functionalist thesis very much more sophisticated. If the analysis more closely represented the details of the discourse of psychologists, it could perhaps be more helpful to them when they need it.

Two such needs can be identified immediately. I claim that strict localization of function is achieved if and only if within some scope one finds a physiological term (or terms) which pick out just those neural arangements subserving a given function. Otherwise functions are only loosely localized. Again, this analysis is somewhat schematic. It does not capture the subtleties of the various arguments over strict or loose localization of function. How is experimental evidence employed in these arguments? What sort of rebuttals can proponents of strict localization offer? What sort of assurance about particular identities would be required if functions were loosely localized? How could one experimentally confirm or falsify particular identities? The various conceptual problems associated with localization of function require further elaboration.

One final area will be mentioned in which more detailed work is required in order to extend the analysis. I argued that what one takes to be function terms or structure terms depends on one's definition of the black box under investigation. In particular, physiological terms in psychological contexts count as structural terms, but when considering neural structures as black boxes, the physiologist often employs functional terms. I also presented a schema which relates psychological and physiological terms in a unified model. This schema does not deal with various details of physiological discourse or the different uses of terms one can find therein. The interrelation between the two classes of terms is more complicated than the analysis suggests, and needs further elaboration.

In all these ways I have presented merely the beginning of an analysis of the relation of psychological terms to physiological terms, and some of the general claims I made will undoubtedly require revision or elaboration. The need for such qualifications is not a cause for despair, but prompts the sobering realization that different domains of discourse can be related in different ways, and the connecting principles required may take different forms. Perhaps the reduction of psychological models of human cognition or semantics would proceed along lines different from those I have sketched here.

Clarifying the features of the particular case is all one can ask, and it is the only sort of clarification that could be of use to the psychologist. The search for all-inclusive analyses of explanatory reduction had best be replaced by attention to the particularities of discourse in different fields.

Bibliography

A. *Physiology*

Akiskal, H. S., and McKinney, W. T., Jr. 'Depressive disorders: toward a unified hypothesis'. *Science,* 182 (1973), 20–9.

Antelman, S. A., *et al.* '6 Hydroxydopamine, noradrenergic reward, and schizophrenia'. *Science*, 175 (1971), 919–23.

Barker, J. L., *et al*. 'Opiate peptide modulation of amino acid responses suggests novel form of neuronal communication'. *Science*, 199 (31 Mar. 1978), 1451–3.

Belluzzi, J. D., Ritter, S., Wise, C. D., and Stein, L. 'Substantia nigra self stimulation: dependence on noradrenergic reward pathways'. *Behavioural Biology,* 13 (1975), 103–11.

Ben Ari, Y., and La Salle, G. Le Gal. 'Plasticity at unitary level II. Modifications during sensory-sensory association procedures'. *Electroencephalography and Clinical Neurophysiology,* 32 (1972), 667–79.

Brady, Joseph. 'Emotion Revisited'. *Journal of Psychiatric Research*, 8 (1971), 363–84.

Butter, C. M., *et al.* Effects of orbital-frontal lesions on aversive and aggressive behaviour in rhesus monkeys'. *Journal of Comparative and Physiological Psychology*, 72 (1970), 132–44.

Butter, C. M. 'Perseveration in extinction and in discrimination reversal tasks following selective frontal ablations in *macaca mulatta'. Physiology and Behaviour*, 4 (1969), 163–71.

Collewijn, H., Winterson, B. J., and Dubois, M. F. W. 'Optokinetic eye movements in albino rabbits: inversion in anterior visual field'. *Science*, 199 (24 Mar. 1978), 1351–3.

Coppen, Alex. 'The biochemistry of affective disorders'. *British Journal of Psychiatry,* 113 (1967), 1237-64.

— 'Indoleamines and affective disorders'. *Journal of Psychiatric Research,* 9 (1972), 163–71.

Eleftheriou, Basil E. (ed.) *Neurobiology of the amygdala.* London: Plenum Press, 1972.

Ervin, F. R., Mark, V. H., and Stevens, J. 'Behavioural and affective responses to brain stimulation in man'. *Neurobiological Aspects of Psychopathology*. Eds. J. Zubin and C. Shagass. New York: Grune and Stratton, 1969.

Fibiger, H. C., Phillips, A. G., and Zis, A. P. 'Deficits in instrumental responding after 6 – hydroxydopamine lesions of the nigraneostriatal dopaminergic projection'. *Pharmacology, Biochemistry and Behaviour,* 2 (1974), 87-96.

Fuster, J. M., and Uyeda, A. A. 'Reactivity of limbic neurons of the monkey to appetitive and aversive signals'. *Electroencephalography and Clinical Neurophysiology*, 30 (1971), 281-93.

Galaburda, A. M., LeMay, M., Kemper, T. L., and Geschwind N. 'Right-left asymmetries in the brain'. *Science*, 199 (24 Feb. 1978), 852-6.

German, D. G., and Bowden, D. M. 'Catecholamine system as the neural substrate for intercranial self stimulation: a hypothesis'. *Brain Research*, 73 (1974), 381-419.

Gloor, P. 'Temporal lobe epilepsy'. *Neurobiology of the Amygdala*. Ed. Basil E. Eleftheriou.

Goldman, P. S., *et al.* 'Analysis of the delayed-alternation deficit produced by dorsolateral prefrontal lesions in the rhesus monkey'. *Journal of Comparative and Physiological Psychology*, 77 (1971), 212-20.

Goldstein, Murray. 'Brain research and violent behaviour'. *Archives of Neurology*, 30 (1974), 1-35.

Grossman, S. P. *Textbook of Physiological Psychology*. New York: John Wiley & Sons, 1967.

Höfelt, T., Ljungdahl, Å., Fuxe, K., and Johansson, O. 'Dopamine nerve terminals in the rat limbic cortex'. *Science*, 184 (1974), 177-9.

Iverson, Susan D., and Iversen, Leslie L. *Behavioural Pharmacology*. Oxford: Oxford University Press, 1975.

Jeeves, M. A. 'Psychological studies of three cases of congenital agnesis of the corpus callosum'. *Functions of the Corpus Callosum*. Ciba Foundation Study Group N. 20. Eds. A. V. de Reuck and Ruth Porter. London: J. & A. Churchill Ltd., 1965.

Jones, B., and Mishkin, M. 'Limbic lesions and the problem of stimulus-reinforcement associations', *Experimental Neurology*, 36 (1972), 362-77.

Jones, E. G., and Powell, T. P. S. 'An anatomical study of converging sensory pathways within the cerebral cortex of the monkey'. *Brain*, 93 (1970), 793-820.

Jouvet, M. 'Neurophysiology of the states of sleep'. *Physiology Review* 47 (1967), 117-77.

Kelly, P. H., Rolls, E. T., and Shaw, S. G. 'Functions of catecholamines in brain-stimulation reward'. *Brain Research*, 66 (1974), 363-4.

Kety, S. S. 'Neurochemical aspects of emotional behaviour'. *Physiological Correlates of Emotion*. Ed. Perry Black. London: Academic Press, 1970.

Klerman, Gerald L. 'Drug therapy of clinical depressions'. *Journal of Psychiatric Research*, 9 (1972), 253-70.

Kraines, S. H. 'Neurological theory of depression'. *American Journal of Psychiatry*, 114 (1957), 206-11.

Le Magnen, Jacques, *et al.* 'Role of a lipostatic mechanism in regulation of feeding of energy balance in rats'. *Journal of Comparative and Physiological Psychology*, 84, 1 (July 1973), 1-23.

Luria, A. R. 'Human brain and psychological processes'. *Mood, States and Mind*. Ed. Karl H. Pribram.

— *The Working Brain.* Harmondsworth, Middx.: Penguin Books, 1973.

MacLean, P. D. 'Psychosomatic disease and the "visceral brain"'. *Psychosomatic Medicine,* 11 (1949) 338-53.

McCleary, Robert A., and Moore, Robert Y. *Subcortical Mechanisms of Behaviour.* London: Basic Books, 1965.

Mark, Vernon H., Ervin, Frank R., and Sweet, William H. 'Deep temporal lobe stimulation in man'. *Neurobiology of the Amygdala.* Ed. Basil E. Eleftheriou.

Marshall, John F., Richardson, J. Steven, and Teitelbaum, Philip. 'Nigro-striatal bundle damage and the lateral hypothalamic syndrome'. *Journal of Comparative and Physiological Psychology,* 87, 5 (Nov. 1974), 808-30.

Melzack, Ronald. *The Puzzle of Pain.* Harmondsworth, Middx.: Penguin Books, 1973.

Milner, Peter M. *Physiological Psychology.* London: Holt, Rinehart & Winston, 1970.

Moyer, K. E. 'A preliminary physiological model of aggressive behaviour'. *The Physiology of Aggression and Defeat,* Eds. Basil E. Eleftheriou and John P. Scott. London: Plenum Press, 1971.

Narabayashi, H. 'Stereotaxic amygdalotomy'. *Neurobiology of the Amygdala.* Ed. Basil E. Eleftheriou.

Nauta, Walle J. H. 'The problem of the frontal lobe: a reinterpretation'. *Journal of Psychiatric Research,* 8 (1971), 167-87.

Neill, D. B., Boggan, W. O., and Grossman, S. P. 'Impairment of avoidance performance by intrastriatal administration of 6 hydroxydopamine'. *Pharmacology, Biochemistry and Behaviour,* 2 (1974), 97-103.

Olds, J. 'Physiological mechanisms of reward'. *Nebraska Symposium on Motivation.* Ed. M. R. Jones. University of Nebraska Press, 1955, pp. 73-134.

Pribram, Karl H. (ed.) *Mood, States and Mind. Brain and Behaviour,* Vol. I. Harmondsworth, Middx.: Penguin Books, 1969.

Randrup, A., *et al.* 'Mania, depression and brain dopamine'. *Current Developments in Psychopharmacology.* Eds. Walter B. Essman and L. Valzelli. Vol II. New York: Spectrum Publications, 1975.

Roberts, W. W. 'Are hypothalamically motivational mechanisms functionally and anatomically specific?: *Brain, Behaviour and Evolution,* 2 (1969), 317-42.

Rolls, E. T. *The Brain and Reward.* Oxford: Pergamon, 1975.

— 'The neural basis of brain stimulation reward'. *Progress in Neurobiology,* 3 (1974), 71-160.

— *et al.* 'The relative attenuation of self-stimulation, eating and drinking produced by dopamine receptor blockade'. *Psychopharmacologia* (Berlin), 38 (1974), 219-30.

Rosvold, H. E. 'The pre-frontal cortex and caudate nucleus: a system for affecting correction in response mechanisms'. *Mind as a Tissue.* Ed. C. Rupp. New York: Hoeber, 1968.

Sanger, F., *et al.* 'Nucleotide sequence of bacteriophage ØX174 DNA'. *Nature,* 265 (24 Feb. 1977), 687–95.

Schildkraut, J. J. 'The catecholamine hypothesis of affective disorders: a review of supporting evidence'. *American Journal of Psychiatry,* 122 (1965), 509–22.

— 'Neuropharmacology of the affective disorders'. *Annual Review of Pharmacology,* 13 (1973), 427–54.

— and Kety, S. S. 'Biogenic amines and emotion'. *Science,* 156 (1967), 21–30.

Sclafani, Anthony, and Kluge, Lawrence. 'Food motivation and body weight levels in hypothalamic hyperphagic rats'. *Journal of Comparative and Physiological Psychology,* 86, 1 (Jan. 1974), 28–46.

Smythies, J. R. *Brain Mechanisms and Behaviour.* Second Edition. Oxford: Blackwell Scientific Publications, 1970.

Snyder, S. H., *et al.* 'Drugs, neurotransmitters and schizophrenia'. *Science,* 184 (1974), 1243–52.

Sokolov, Ye. N. 'Higher nervous functions: the orienting reflex'. *Annual Review of Physiology,* 25 (1963), 545–80.

Stein, Larry. 'Neurochemistry of reward and punishment: some implications for the etiology of schizophrenia'. *Journal of Psychiatric Research,* 8 (1971), 345–61.

— Belluzzi, J. D., and Wise, C. David. 'Norepinephrine self stimulation pathways: implications for long-term memory and schizophrenia'. *Proceedings of the First International Conference on Intra Cranial Self Stimulation.* Ed. E. T. Rolls, 1976 (in press).

— and Wise, C. David. 'Amphetamine and noradrenergic reward pathways'. *Frontiers in Catecholamine Research.* Eds. E. Usdin and S. H. Snyder. Oxford: Pergamon, 1973.

— — 'Possible etiology of schizophrenia'. *Science,* 171 (1971), 1032–6.

Tannenbaum, Gloria A., Paxinos, George, and Bindra, Dalbir. 'Metabolic and endocrine aspects of the ventromedial hypothalamic syndrome in the rat'. *Journal of Comparative and Physiological Psychology,* 86, 34 (Mar. 1974), 404–13.

Teuber, H. L. 'Unity and diversity of frontal lobe functions'. *Acta Neurobiologica Experimentalis Warsaw*, 32, 615–56.

Thompson, J. B., and Schwartzbaum, J. S. 'Discrimination behaviour and conditioned suppression (CER) following localized lesion in the amygdala and putamen'. *Psychology Reports,* 15 (1964), 587–606.

Thompson, Richard F. *Foundations of Physiological Psychology,* New York: Harper & Row, 1967.

Toates, F. M., and Oatley, K. 'Computer simulation of thirst and water balance'. *Medical and Biological Engineering,* 8 (1970), 71–87.

Valenstein, E. S. *Brain Control.* New York: John Wiley & Sons, 1973.

— Cox, V. C., and Kakolewski, J. W. 'Reexamination of the role of the hypothalamus in motivation'. *Psychological Review,* 77 (1970), 16–31.

Wise, C. David, and Stein, Larry. 'Dopamine β hydroxylase deficits in the brains of schizophrenic patients'. *Science,* 181 (1973), 344.

— — 'Facilitation of brain self-stimulation by central administration of norepinephrine'. *Science,* 163 (1969), 299-301.

Zbrozyna, A. W. 'The organization of the defence reaction elicited from the amygdala and its connections'. *Neurobiology of the Amygdala.* Ed. Basil E. Eleftheriou.

B. *Psychology*

Anderson, Barry F. *The Psychology Experiment.* Belmont, California: Wadsworth Publishing Company, 1967.

Andrew, A. M. 'Learning Systems'. *Automaton Theory and Learning Systems.* Ed. D. J. Stewart.

Angyal, Andras. *Foundations for a Science of Personality.* New York: The Commonwealth Fund, 1941.

Apter, Michael J. *The Computer Simulation of Behaviour.* London: Hutchinson University Library, 1970.

Arnold, Magda B. (ed.) *The Nature of Emotion.* Harmondsworth, Middx.: Penguin Books, 1968.

Ashby, W. R. 'Principles of the self-organizing system'. *Principles of Self Organization.* Eds. Heinz von Foerster and George W. Zopf. Jr.

— 'Self regulation and requisite variety'. *Systems Thinking.* Ed. F. E. Emery.

Attneave, Fred. *Applications of Information Theory to Psychology.* New York: Holt, Rinehart & Winston, 1959.

Bindra, Dalbir. 'Drive and incentive-motivation'. *Motivation.* Eds. Dalbir Bindra and Jane Stewart.

— 'The inter-related mechanisms of reinforcement and motivation, and the nature of their influence on response'. *Motivation.* Eds. Dalbir Bindra and Jane Stewart.

— and Stewart, Jane (eds.) *Motivation.* Second Edition. Harmondsworth, Middx.: Penguin Books, 1971.

Black, R. W. 'Theories of reward'. *Motivation.* Eds. Dalbir Bindra and Jane Stewart.

Bolles, R. C. *Theory of Motivation.* New York: Harper & Row, 1967.

Bonime, Walter. 'The psychodynamics of neurotic depression'. *American Handbook of Psychiatry.* Ed. Silvano Arieti. Vol III. New York: Basic Books, 1966, 239-55.

Broadbent, D. E. *Behaviour.* London: Eyre & Spottiswoode, 1961.

Clare, Anthony. *Psychiatry in Dissent.* London: Tavistock Publications Ltd., 1976.

Cooper, David. *Psychiatry and Anti-Psychiatry.* Frogmore, St. Albans, Herts.: Paladin, 1970.

Dawkins, Richard. *The Selfish Gene.* New York: Oxford University Press, 1976.

Deutsch, J. A. *The Structural Basis of Behaviour.* Cambridge: Cambridge University Press, 1964.

Duffy, E. 'An explanation of "emotional" phenomena without the use of the concept "emotion"'. *The Nature of Emotion*. Ed. Magda B. Arnold.

Emery, F. E. (ed.) *Systems Thinking*. Harmondsworth, Middx.: Penguin Books, 1969.

George, F. H. *The Brain as a Computer*. Oxford: Pergamon, 1961.

Glickman, S. E. and Schiff, B. B. 'A theory of reinforcement'. *Motivation*. Eds. Dalbir Bindra and Jane Stewart.

Gray, Jeffrey A. *Elements of a Two-Process Theory of Learning*. London: Academic Press, 1975.

— *The Psychology of Fear and Stress*. London: Weidenfeld & Nicolson, 1971.

— 'The structure of the emotions and the limbic system: a theoretical model'. *Physiology, Emotion and Psychosomatic Illness*. Eds. R. Porter and J. Knight. Ciba Foundation Symposium No. 8. New Series. Amsterdam: Associated Scientific Publishers, 1972, pp. 87-120.

Grinker, Roy R., and Nunnally, Jum C. 'The phenomena of depressions'. *Psychopathology Today*. Ed. William S. Sahakian. Itasca, Illinois: F. E. Peacock Publishers, 1970.

Hebb, D. O. 'Drives and the CNS (conceptual nervous system)'. *Psychological Review*, 62 (1955), 243-54.

— *Organization of Behavior*. New York: John Wiley & Sons, 1949.

Hilgard, Ernest R. *Theories of Learning*. New York: Appleton-Century-Crofts, 1948.

Hinde, R. A. 'Critique of energy models of instinctive actions'. *Motivation*. Eds. Dalbir Bindra and Jane Stewart.

Hull, Clark L. 'Drive as an intervening variable in a formal behaviour system'. *Motivation*. Eds. Dalbir Bindra and Jane Stewart.

— 'Law of primary reinforcement'. *Motivation*. Eds. Dalbir Bindra and Jane Stewart.

— 'Principles of behavior'. *Motivation*. Eds. Dalbir Bindra and Jane Stewart.

— *Principles of Behavior*. New York: D. Appleton-Century, 1943.

Johnson, Homer H., and Solso, Robert L. *An Introduction to Experimental Design in Psychology: A Case Approach*. New York: Harper & Row, 1971.

Kiloh, L. G., and Garside, R. F. 'Distinction between neurotic and endogenous depression'. *British Journal of Psychiatry*, 109 (1963), 451-63.

Kremyanskiy, V. I. 'Certain peculiarities of organisms as a "system" from the point of view of physics, cybernetics, and biology'. *Systems Thinking*. Ed. F. E. Emery.

Lashley, K. S. 'Drive as facilitation of specific neural mechanisms'. *Motivation*. Eds. Dalbir Bindra and Jane Stewart.

Leeper, R. W. 'A motivational theory of emotion to replace "emotion as disorganized response"'. *Adaptation*. Ed. Karl H. Pribram.

Lorenz, Konrad. 'The comparative method in studying innate behaviour patterns'. *Symposia of the Society for Experimental Biology*, 4 (1950), 221-68.

— 'Innate bases of learning'. *On the Biology of Learning.* Ed. Karl H. Pribram. New York: Harcourt, Brace & World, 1969.

— *On Agression.* New York: Harcourt, Brace & World, 1966.

MacFarland, D. J. *Feedback Mechanisms in Animal Behaviour.* London: Academic Press, 1971.

Miller, N. E. 'Some reflections on the law of effect produce a new alternative to drive reduction'. *Nebraska Symposium on Motivation.* Ed. M. R. Jones. University of Nebraska Press, 1963, pp. 65-107.

Milsun, John H. *Biological Control Systems Analysis.* New York: McGraw-Hill, 1966.

Morgan, C. T. 'A theory of motivation'. *Motivation.* Eds. Dalbir Bindra and Jane Stewart.

Mowrer, O. Hobart. *Learning Theory and Behavior.* New York: John Wiley & Sons, 1960.

Plutchik, Robert. *Foundations of Experimental Research.* New York: Harper & Row, 1968.

Pribram, Karl H. (ed.) *Adaptation. Brain and Behaviour,* Vol. IV. Harmondsworth, Middx.: Penguin Books, 1969.

— *Languages of the Brain.* Englewood Cliffs, N.J.: Prentice-Hall, 1971.

Rose, Steven. *The Conscious Brain.* Revised Edition. Harmondsworth, Middx.: Penguin Books, 1976.

Sacks, Oliver. *Awakenings.* Harmondsworth, Middx.: Penguin Books, 1976.

Seligman, Martin E. P. *Helplessness.* San Francisco: W. H. Freeman & Co., 1975.

Simon, H. A. 'Rational choice and the structure of the environment'. *Systems Thinking.* Ed. F. E. Emery.

Skinner, B. F. *Science and Human Behaviour.* New York: Macmillan Co., 1953.

— 'Why is a reinforcer reinforcing?'. *Motivation.* Eds. Dalbir Bindra and Jane Stewart.

Sommerhoff, Gerd. 'The abstract characteristics of living systems'. *Systems Thinking.* Ed. F. E. Emery.

— *Logic of the Living Brain.* London: John Wiley & Sons, 1974.

Spence, K. W. 'Incentive motivation'. *Motivation.* Eds. Dalbir Bindra and Jane Stewart.

Sperry, R. W. 'Orderly function with disorderly structure'. *Principles of Self Organization.* Eds. Heinz von Foerster and George W. Zopf Jr.

Stellar, Eliot. 'The physiology of motivation'. *Psychological Review,* 61 (1954), 5-22.

Stewart, D. J. (ed.) *Automaton Theory and Learning Systems.* London: Academic Press, 1967.

— 'Logical nets and organisms that learn'. *Automaton Theory and Learning Systems.* Ed. D. J. Stewart.

Strongman, K. T. *Psychology of Emotion.* London: John Wiley & Sons, 1973.

Thorndike, E. L. 'The Law of effect'. *Motivation.* Eds. Dalbir Bindra and Jane Stewart.

Tinbergen, N. 'Hierarchical organization of instinctive actions'. *Motivation.* Eds. Dalbir Bindra and Jane Stewart.

Valle, Fred P. *Motivation: Theories and Issues.* Monterey, California: Brooks/Cole Publishing Company, 1975.

von Bertalanffy, Ludwig. *General System Theory.* Harmondsworth, Middx.: Penguin Books, 1973.

— 'The theory of open systems in physics and biology'. *Systems Thinking.* Ed. F. E. Emery.

von Foerster, Heinz and Zopf, George W. Jr. (eds.) *Principles of Self Organization.* Oxford: Pergamon, 1962.

Young, P. T. 'Affective processes'. *The Nature of Emotion.* Ed. Magda B. Arnold.

C. *Philosophy*

Anscombe, G. E. M. *Intention.* Second Edition. Oxford: Basil Blackwell, 1972.

Armstrong, D. M. *Bodily Sensations.* London: Routledge & Kegan Paul, 1969.

Armstrong, D. M. 'The nature of mind'. *The Mind-Brain Identity Theory.* Ed. C. V. Borst.

Baier, Kurt. 'Smart on sensations'. *The Mind-Brain Identity Theory.* Ed. C. V. Borst.

Bedford, Errol. 'Emotions'. *The Philosophy of Mind.* Ed. V. C. Chappell. Englewood Cliffs, Prentice-Hall, 1962.

Bergmann, Gustav. 'Purpose, function, scientific explanation'. *Readings in the Philosophy of the Social Sciences.* Ed. May Brodback.

Berlin, Isaiah. 'Historical inevitability'. *Four Essays on Liberty.* By Isaiah Berlin. Oxford: Oxford University Press, 1969.

Berofsky, Bernard (ed.) *Free Will and Determinism.* London: Harper & Row, 1966.

Bhaskar, Roy. *A Realist Theory of Science.* Leeds: Leeds University Press, 1975.

Boden, Margaret A. *Purposive Explanation in Psychology.* Cambridge, Mass.: Harvard University Press, 1972.

Borger, Robert, and Cioffi, Frank (eds.) *Explanation in the Behavioural Sciences.* Cambridge: Cambridge University Press, 1970.

Borst, C. V. (ed.) *The Mind-Brain Identity Theory.* London: Macmillan, 1970.

Braithwaite, R. B. 'Models in the empirical sciences'. *Readings in the Philosophy of Science.* Ed. Baruch A. Brody.

— *Scientific Explanation.* Cambridge: Cambridge University Press, 1959.

Brandt, Richard B. 'Doubts about the identity theory'. *Dimensions of Mind.* Ed. Sidney Hook. New York: New York University Press, 1960.

Brandt, R., and Kim, J. 'The logic of the identity theory'. *Journal of Philosophy,* 64 (1967), 515-37.

Brittan, C. G. 'Explanation and reduction'. *Journal of Philosophy,* LXVII, 13 (9 July 1970), 446-57.

Brodbeck, May (ed.) *Readings in the Philosophy of the Social Sciences.* New York: Macmillan Co., 1968.

Brody, Baruch A. (ed.) *Readings in the Philosophy of Science.* Englewood Cliffs, N.J.: Prentice-Hall, 1970.

Bromberger, Sylvain. 'An approach to explanation'. *Analytical Philosophy.* Ed. J. R. Butler. Second Series. Oxford: Basil Blackwell, 1965.

Buck, R. C. 'On the logic of general systems theory'. *Minnesota Studies in the Philosophy of Science.* Vol. I. Eds. Herbert Feigl and Grover Maxwell. Minneapolis: University of Minnesota Press, 1956.

Carnap, Rudolf. *Foundations of Logic and Mathematics.* International Encyclopedia of Unified Science, Vol. I, no. 3. Chicago: University of Chicago Press, 1939.

— *An Introduction to the Philosophy of Science.* Ed. Martin Gardner. New York: Basic Books, 1966.

— 'The methodological character of theoretical concepts'. *Minnesota Studies in the Philosophy of Science.* Vol. I. Eds. Herbert Feigl and Grover Maxwell. Minneapolis: University of Minnesota Press, 1956.

— 'Theories as partially interpreted formal systems'. *Readings in the Philosophy of Science.* Ed. Baruch A. Brody.

Causey, R. 'Attribute identities in micro-reductions'. *Journal of Philosophy,* LXIX, 14 (3 Aug. 1972), 407-22.

Cornman, J. 'The identity of mind and body'. *The Mind-Brain Identity Theory.* Ed. C. V. Borst.

— *Materialism and Sensations.* New Haven, Connecticut: Yale University Press, 1971.

— 'On the elimination of "sensations" and sensations'. *Modern Materialism.* Ed. John O'Connor.

Davidson, Donald. 'Actions, reasons, and causes'. *Free Will and Determinism.* Ed. Bernard Berofsky.

— 'Mental events'. *Experience and Theory.* Eds. Lawrence Foster and J. W. Swanson. Amherst: University of Massachusetts Press, 1970.

— 'Psychology as philosophy'. *Philosophy of Psychology.* Ed. S. C. Brown. London: Macmillan, 1974.

Dennett, D. C. *Content and Consciousness.* London: Routledge & Kegan Paul, 1969.

— 'Intentional systems'. *Journal of Philosophy,* LXVIII, 4 (25 Feb. 1971), 87-106.

Emery, F. E. (ed.) *Systems Thinking.* Harmondsworth, Middx.: Penguin Books, 1969.

Enç, Berent. 'Identity statements and micro-reductions'. *Journal of Philosophy,* LXXIII, 11 (10 June 1976), 285-305.

Farrell, B. A. 'Experience'. *Mind,* LIX (1950), 170-98.

Fay, Brian. *Social Theory and Political Practice.* London: George Allen & Unwin, 1975.

Feigl, Herbert. *The 'Mental' and the 'Physical': The Essay & a Postscript.* Minneapolis: University of Minnesota Press, 1967.

— 'Mind-body, *not* a pseudo-problem'. *The Mind-Brain Identity Theory.* Ed. C. V. Borst.

— 'Physicalism, unity of science, and the foundations of psychology'. *The Philosophy of Rudolf Carnap.* Ed. P. A. Schilpp. Lasalle, Illinois: Open Court Publishing Co., 1963.

Feyerabend, Paul. 'Comment: "Mental events and the brain"'. *The Mind-Brain Identity Theory.* Ed. C. V. Borst.

— 'Explanation, reduction, and empiricism'. *Minnesota Studies in the Philosophy of Science.* Vol. III. Eds. Herbert Feigl and Grover Maxwell. Minneapolis: University of Minnesota Press, 1962.

— 'Materialism and the mind-body problem'. *The Mind-Brain Identity Theory.* Ed. C. V. Borst.

Fodor, Jerry A. 'Functional explanation in psychology'. *Readings in the Philosophy of the Social Sciences.* Ed. May Brodbeck.

— *The Language of Thought.* Hassocks, Sussex: The Harvester Press, 1976.

— *Psychological Explanation.* New York: Random House, 1968.

Habermas, Jurgen. 'Technology and science as "ideology"'. *Toward a Rational Society.* By Jurgen Habermas. Trans. by Jeremy J. Shapiro. Boston: Beacon Press, 1970.

Hamlyn, D. W. 'Behaviour'. *Philosophy,* XXVIII (1953), 132–45.

Harman, Gilbert. *Thought.* Princeton: Princeton University Press, 1973.

Harré, Romano. *The Principles of Scientific Thinking.* London: Macmillan, 1970.

— *Theories and Things.* London: Sheed and Ward, 1961.

— and Secord, P. F. *The Explanation of Social Behaviour.* Oxford: Basil Blackwell, 1972.

Hempel, Carl G. 'Aspects of scientific explanation'. *Aspects of Scientific Explanation and Other Essays in the Philosophy of Science.* New York: The Free Press, 1965.

— 'A logical appraisal of operationism'. *Aspects of Scientific Explanation.*

— 'The theoretician's dilemma'. *Aspects of Scientific Explanation.*

— 'The logic of functional analysis'. *Readings in the Philosophy of the Social Sciences.* Ed. May Brodbeck.

— *Philosophy of Natural Science.* Englewood Cliffs, N.J.: Prentice-Hall, 1966.

— 'Reduction: ontological and linguistic facets'. *Philosophy, Science, and Method.* Eds. S. Morgenbesser, P. Suppes, and M. White. New York: St. Martin's Press, 1969.

Hook, Sidney. *Dimensions of Mind.* New York: New York University Press, 1960.

Kalke, William. 'What is wrong with functionalism?'. *Nous,* III, 1 (Feb. 1969), 83–94.

Kemeny, John G., and Oppenheim, Paul. 'On reduction'. *Readings in the Philosophy of Science.* Ed. Baruch A. Brody.

Kim, Jaegwon. 'On the psycho-physical identity theory'. *American Philosophical Quarterly,* 3 (1966), 227–35.

— 'Psycho-physical correlation laws and theories of mind'. *Theoria,* XXXIII (1967), 198–210.

Kripke, Saul. 'Identity and necessity'. *Identity and Individuation.* Ed. Milton K. Munitz. New York: New York University Press, 1971.

Kuhn, Thomas. *The Structure of Scientific Revolutions.* Second Edition. Chicago: University of Chicago Press, 1970.

Laslett, Peter (ed.) *The Physical Basis of Mind.* New York: Macmillan Co., 1950.

Lewis, David. 'An argument for the identity theory'. *Journal of Philosophy,* LXIII, 1 (6 Jan. 1966) 17–25.

— 'Psychological and theoretical identifications'. *Australasian Journal of Philosophy,* L, 3 (Dec. 1972), 249–58.

Louch, A. R. *Explanation and Human Action.* Berkeley: University of California Press, 1966.

Mackie, J. L. *The Cement of the Universe.* Oxford: Clarendon Press, 1974.

Marcuse, Herbert. 'Technological rationality and the logic of domination'. *One Dimensional Man.* By Herbert Marcuse. London: Abacus Books Ltd., 1972.

Mates, Benson. *Elementary Logic.* Second Edition. New York: Oxford University Press, 1972.

Maxwell, Grover. 'The ontological status of theoretical entities'. *Minnesota Studies in the Philosophy of Science.* Vol. III. Eds. Herbert Feigl and Grover Maxwell. Minneapolis: University of Minnesota Press, 1962.

Melden, A. I. *Free Action.* London: Routledge & Kegan Paul, 1961.

Nagel, Ernest. 'A formalization of functionalism'. *Logic without Metaphysics.* By Ernest Nagel. Glencoe, Illinois: The Free Press, 1956.

— 'Issues in the logic of reductive explanations'. *Mind, Science and History.* Eds. H. Keifer and M. Munitz. Albany: State University of New York Press, 1970, pp. 117–38.

— *The Structure of Science.* London: Routledge & Kegan Paul, 1961.

Nagel, Thomas. 'Physicalism'. *The Mind-Brain Identity Theory.* Ed. C. V. Borst.

Nelson, R. J. 'Mechanism, functionalism, and the identity theory'. *Journal of Philosophy,* LXXIII, 13 (15 July 1976), 365–85.

Nickles, T. 'Two concepts of inter-theoretic reduction'. *Journal of Philosophy,* LXX, 7 (12 Apr. 1973), 181–201.

O'Connor, John. *Modern Materialism: Readings on Mind-Body Identity.* New York: Harcourt, Brace & World, 1969.

Oppenheim, Paul, and Putnam, Hilary. 'Unity of science as a working hypothesis'. *Minnesota Studies in the Philosophy of Science.* Vol. II. Eds. Herbert Feigl and Grover Maxwell. Minneapolis: University of Minnesota Press, 1958.

Peters, R. S. *The Concept of Motivation.* London: Routledge & Kegan Paul, 1960.

Place, U. T. 'Is consciousness a brain process?'. *The Mind-Brain Identity Theory.* Ed. C. V. Borst.

— 'Materialism as a scientific hypothesis'. *The Mind-Brain Identity Theory.* Ed. C. V. Borst.

Popper, Karl R. *Conjectures and Refutations.* New York: Harper Torchbooks, 1965.

Putnam, Hilary. 'Brains and behaviour'. *Mind, Language and Reality.* Philosophical Papers, Vol. II. By Hilary Putnam. Cambridge: Cambridge University Press, 1975.

— 'Language and philosophy'. *Mind, Language and Reality.* By Hilary Putnam.

— 'The mental life of some machines'. *Mind, Language and Reality.* By Hilary Putnam.

— 'Minds and machines'. *Mind, Language and Reality.* By Hilary Putnam.

— 'The nature of mental states'. *Mind, Language and Reality.* By Hilary Putnam.

— 'Philosophy and our mental life'. *Mind, Language and Reality.* By Hilary Putnam.

— 'Robots: machines or artificially created life?'. *Mind, Language and Reality.* By Hilary Putnam.

— 'Is semantics possible?'. *Language, Belief and Metaphysics.* Eds. H. E. Kiefer and M. K. Munitz. Albany: State University of New York Press, 1970.

Quine, W. V. 'On mental entities'. *Modern Materialism.* Ed. John O'Connor.

— *Ontological Relativity and Other Essays.* New York: Columbia University Press, 1969.

Rorty, Richard. 'In defence of eliminative materialism'. *Review of Metaphysics,* XXIV, 1 (Sept. 1970), 112–21.

— 'Functionalism, machines, and incorrigibility'. *Journal of Philosophy,* LXIX, 8 (20 Apr. 1972), 203–20.

— 'Mind-body identity, privacy and categories'. *The Mind-Brain Identity Theory.* Ed. C. V. Borst.

Rosenthal, David M. 'Mentality and neutrality'. *Journal of Philosophy,* LXXIII, 13 (15 July 1976), 386–415.

Ryle, Gilbert. *The Concept of Mind.* New York: Barnes & Noble, 1949.

Sayre, Kenneth. *Cybernetics and The Philosophy of Mind.* London: Routledge & Kegan Paul, 1976.

Schaffner, Kenneth F. 'Approaches to reduction'. *Philosophy of Science,* XXXIV, 2 (June 1967), 137–47.

Schützenberger, M. P. 'A tentative classification of goal-seeking behaviours'. *Systems Thinking.* Ed. F. E. Emery.

Sellars, Wilfrid. 'The identity approach to the mind-body problem'. *Modern Materialism.* Ed. John O'Connor.

— 'The language of theories'. *Readings in the Philosophy of Science.* Ed. Baruch A. Brody.

— *Science, Perception and Reality*. London: Routledge & Kegan Paul, 1961.

Shaffer, Jerome. 'Could mental states be brain processes?'. *The Mind-Brain Identity Theory*. Ed. C. V. Borst.

— 'Mental events and the brain'. *The Mind-Brain Identity Theory*. Ed. C. V. Borst.

— 'Recent work on the mind-body problem'. *American Philosophical Quarterly*, II (Apr. 1965), 81–104.

Skinner, B. F. *Beyond Freedom and Dignity*. Harmondsworth, Middx.: Penguin Books, 1971.

Sklar, Lawrence. 'Types of inter-theoretic reduction'. *British Journal for the Philosophy of Science*, XVIII, 2 (Aug. 1967), 109–24.

Smart, J. J. C. 'Brain processes and incorrigibility'. *The Mind-Brain Identity Theory*. Ed. C. V. Borst.

— 'Materialism'. *The Mind-Brain Identity Theory*. Ed. C. V. Borst.

— *Philosophy and Scientific Realism*. London: Routledge & Kegan Paul, 1963.

— 'Sensations and brain processes'. *The Mind-Brain Identity Theory*. Ed. C. V. Borst.

Spector, Marshall. 'Models and theories'. *Readings in the Philosophy of Science*. Ed. Baruch A. Brody.

Stevenson, J. T. '"Sensations and brain processes": a reply to J. J. C. Smart'. *The Mind-Brain Identity Theory*. Ed. C. V. Borst.

Sutherland, N. S. 'Is the brain a physical system?'. *Explanation in the Behavioural Sciences*. Eds. Robert Borger and Frank Cioffi.

Taylor, Charles. *The Explanation of Behaviour*. London: Routledge & Kegan Paul, 1964.

— 'The explanation of purposive behaviour'. *Explanation in the Behavioural Sciences*. Eds. Robert Borger and Frank Cioffi.

— 'Mind-body identity, a side issue?'. *The Mind-Brain Identity Theory*. Ed. C. V. Borst.

Toulmin, Stephen. 'Reasons and causes'. *Explanation in the Behavioural Sciences*. Eds. Robert Borger and Frank Cioffi.

Vesey, G. N. A. *The Embodied Mind*. London: George Allen and Unwin, 1965.

von Wright, George Henrik. *Explanation and Understanding*. London: Routledge & Kegan Paul, 1971.

Winch, Peter. *The Idea of a Social Science and its Relation to Philosophy*. London: Routledge & Kegan Paul, 1963.

Wisdom, John. *Other Minds*. Los Angeles: University of California Press, 1968.

Woodfield, Andrew. *Teleology*. Cambridge: Cambridge University Press, 1976.

Zaffron, R. 'Identity, subsumption, and scientific explanation'. *Journal of Philosophy*, LXVIII, 23 (2 Dec. 1971), 849–60.

Index